RESCUING SCIENCE

RESCUING SCIENCE

Restoring Trust in an Age of Doubt

Paul M. Sutter

ROWMAN & LITTLEFIELD

Lanham • Boulder • New York • London

Published by Rowman & Littlefield
An imprint of The Rowman & Littlefield Publishing Group, Inc.
4501 Forbes Boulevard, Suite 200, Lanham, Maryland 20706
www.rowman.com

86-90 Paul Street, London EC2A 4NE

British Library Cataloguing in Publication Information Available

Library of Congress Cataloging-in-Publication Data Available

ISBN: 978-1-5381-8161-4 (cloth : alk. paper)
ISBN: 978-1-5381-8162-1 (ebook)

∞™ The paper used in this publication meets the minimum requirements of American National Standard for Information Sciences—Permanence of Paper for Printed Library Materials, ANSI/NISO Z39.48-1992.

For Kate

For your integrity
For your wisdom
For your love

Contents

Prologue
The Modern Science

THIS IS PROBABLY THE MOST difficult book I've ever written. And I've written about *quantum mechanics*, so that's telling you something. The following pages will be filled with words I'm going to struggle typing, for several reasons. For starters, I absolutely love science. That may not seem like that big of a difficulty, but it is when you're trying to write a book . . . criticizing science.

But—and this will be a recurring theme throughout this book—while I love the methods of science and the beautiful way it reveals the natural world before our very eyes (or instruments), throughout my career I have witnessed consistent and systematic failures of the *practice* of science. I believe that modern science is sick. Ill. Unhealthy. I believe that science is failing the public, failing the next generation of scientists, and failing itself. I believe that the way we do science today is so much less than what it could be.

Honestly, this book is a love letter to science. Just not one of those passive-aggressive love letters where you highlight all the flaws of your companion in contrast to your own strengths. No, it's a love letter to what I want science to be. What I hope it can be. What I honestly believe science is capable of in our modern society.

I believe that most scientists, like most people, are fundamentally good people (you may call me naïve for that, but I feel it's a fair way to live—after all, I'd prefer it if strangers assumed that I was a decent person). I believe that they are tremendously gifted thinkers and dedicated hard workers. I believe that science is making the world a better place, both through the technological benefits of a greater understanding of our universe and through the pure gift of knowledge of that universe.

But I also believe that the way science is currently practiced is harming its relationship to the wider world and ultimately undermining its own future. I believe—and hopefully will argue successfully throughout this book—that scientists are making critical mistakes in their relationships to each other, to their younger colleagues, and to nonscientists.

I believe that the institution of science has grown in strange directions since its inception a few centuries ago. While the roots of the scientific method stretch back across thousands of years and can be found in dozens of cultures, what we call "science" really got its start in the late sixteenth century with the realization, among other things, that perhaps the Earth was not at the center of the universe. Since then, the scientific method has grown and evolved, becoming an independent line of inquiry in its own right, separate from philosophy, theology, and mathematics (although there's still a fair bit of math and philosophy involved, but we'll get to that).

But largely through its history, science was a game played by rich European aristocrats, something to pass the time, like painting or music. It was pretty fashionable to have, say, a giant telescope on the grounds of your estate. Over the course of the nineteenth century, however, "scientist" became a job description (and the word "scientist" was coined)—it became a *thing* that you could pursue as a career. Still, the vast majority of people didn't interact with scientists, or science itself, in any meaningful way.

In the late nineteenth and early twentieth century, multiple revolutions in our understanding of the world around us (evolution, germ theory, quantum mechanics, and the rest), along with associated technological advancement, catapulted science to the forefront of the world stage. In the middle of the last century, this growth accelerated in the post–World War II boom, with many scientists serving the needs of the nation during the Cold War and the Space Race. As we'll detail in forthcoming chapters, the number of scientists has exploded in recent decades. Scientists became woven into the fabric of society, not just discovering new knowledge but becoming sources of reliable, trusted information. Universities, once the realm of only the most esoteric and exclusive seekers of knowledge, became The Business that put scientists to work, tying the successful education of our population to the scientific output of their researchers.

Science got big.

And in the process of getting big, science molded itself to fit the funding and education requirements of the modern era and contorted itself to meet the expectations of the people paying the bills. I believe that certain traditions, norms, and patterns have not only established themselves in the practice of science but ossified to a dangerous degree, making the scientific apparatus more fragile than it appears.

As we'll explore, I believe that the modern practice of science has become obsessed with money. That it permits fraud and lies in a blind pursuit of publishing papers. That it stifles creativity and new ideas. That it has structured itself so that young scientists have little hope of advancement in a system that exploits their vulnerability. I believe that science has been closed to a diversity of people for far too long. I believe that science has gotten lazy and afraid of exploring risky avenues of research for fear of losing funding. I believe that scientists have disparaged other valid forms of inquiry and closed themselves off to the public. I believe that scientists have demanded respect without earning it and have insisted that the public and policymakers hear what they have to say without warranting a seat at the table.

I believe that science is sick.

I believe that scientists have ignored these issues, and many more, for far too long. That scientists have simply assumed that they are supported, appreciated, and respected—without bothering to do the scientific thing and actually check if that's still true.

The reason I felt compelled to write this book is the same reason that I've felt compelled to write any of my other books: something in my mind and heart is urging me to put pen to paper (or rather finger to keyboard) to tell this story. In this book, you're going to learn a lot about me, personally. I hope you don't mind. I'll be drawing from my own experiences in academia as a scientist. Throughout my career, I've witnessed many aspects of the practice of science that have made me furrow my brow—some things I've seen and heard (and done) just haven't felt right in my gut. Years later, with the benefit of hindsight and some healthy distance between me and the academic world, I've realized that my gut was trying to tell me something (besides the usual moans of hunger), and with further digging and research I discovered that I wasn't alone. What I personally saw as troubling facets turned out to be deep-seated patterns.

And so: this book. A book that will, at first glance, make science look really, really bad. Which is unfortunate because, like I said, I love this stuff. I love science so much that I've spent the vast majority of my professional career being a scientist. My progress through academia was not all that atypical. I got a bachelor's in physics. I got accepted into grad school. I got a PhD in physics. I held a couple postdoctoral research positions. I had offices in universities. I wrote papers. I won grants. I worked with collaborators. I wrote emails—a lot of them. I've given hundreds of talks, visited departments, and worked with colleagues all around the world. I've had many beautiful, close relationships with people I respect and admire.

But this is not a book extolling the virtues of science—there are plenty of those already. This is not the book to tell you how the scientific method is so powerful, how it's opened up vistas great and small, how it's utterly transformed—and continues to transform—the world we live in and how we see it. This is not the book to talk about the scientific method, how we've concocted this precise *machine* for answering questions about the natural workings of the universe. Science is indeed potent. Science is revelatory. Science is transformative. Science is worthy of respect . . . and even love. Scientists themselves are passionate, caring, driven people. The world opened up by science is beautiful, transcendent, even magical. But that ain't this book.

This is the book that explores the darker side of science. The mistakes and follies conducted and permitted by scientists. The flaws. The warts. The sickness. I have no issues with science itself and how its methodology is organized, but I am troubled by the *ways* that science is practiced in the modern world, and worried for its future.

As we'll see, the flaws that we will explore in the practice of science are not unique to science itself. Science is, after all, performed by humans—the very same humans who tend to make the same kinds of mistakes every time they try to organize themselves and get something done.

Most importantly, through this book I am going to make a case of the fundamental sickness in science: that the relationship between science and society is breaking (and in some cases already broken). Between you and me, the last part of the book that I wrote is this very prologue, and looking back over the rest of the book that I'd already written I found myself somewhat surprised with how . . . strong . . . some of my language is about to get. But like I said, what I'm about to explore are issues that I've personally encountered, and like seeing a good friend fall to drug addiction, it's *hard*. I'm going to be tossing out some pretty heavy words: lazy, apathetic, arrogant, prideful, greedy, and more. If you yourself are an academic or a fan of science, those words may be hard to swallow. They were hard to type. But I believe that they are very, very appropriate.

I'm also going to dig deep into the *whys* of how science became sick. The causes, the motivations, the histories. I'll also put forward ideas—sometimes radical—on how I think we can best change course before we hit the metaphorical iceberg. I need to note, however, that every single subject that I bring up in these individual chapters is much, much larger than can be captured in a few tens of thousands of words. Helpfully, I've sprinkled notes and references throughout—I highly encourage you to read those sources, as I hope you'll find them as illuminating (and sometimes motivatingly depressing) as I did. I'm also perfectly aware that my suggestions for solutions barely scratch the surface of what can be done and don't go nearly deep enough into the actual mechanics of change—no single book could really do that, as it's the work of generations.

Almost all of the issues that I'll be highlighting in this book are much bigger than any individual scientist—they are widespread, deep-rooted, systemic issues. But I believe that the root cause of the sickness in science is the millions upon millions of singular choices that individual scientists make, and so I'm going to focus on those choices and those decisions, and what scientists choose (and choose not) to do.

The point of this book is to get you thinking—correction, to get *us* thinking. To be aware of our shortcomings. To recognize the failings and failures for what they are—manifestations of a flawed system—so that we can continue to celebrate the good that science is in our world and strive together to bring about a better future.

Of course, what you're about to read is not an indictment of every single scientist on the face of the earth. There are many good-hearted people in the sciences—including many I've been fortunate enough to work with—who do not participate in some of the vices that I'm going to describe and, even better, are actively trying to mitigate them or eliminate them altogether. But if I had to guess, the vast majority of scientists participate unknowingly in perpetuating these harmful behaviors, not because they actively choose to but because there's simply no other avenue available to them. As we'll see time and time again, scientists are often forced to go against their own good judgment because if they don't, they suffer: they lose access to grants, they get denied promotion, they get kicked out of science altogether. So I don't recommend walking up to the nearest scientist and slapping them in the face for continuing these destructive patterns—it's not necessarily their individual fault.

And yet, as we've seen time and time again throughout history, the act of turning a blind eye, of allowing things to get worse, and of participating in a broken system when you're fully aware of the long-term harm it causes is its own sort of active participation in the charade. Every time a scientist chooses to play the game, they are allowing the system to continue through another generation, widening the cracks in the foundation and entrenching the divides between themselves and society.

So maybe it is their fault. Still, you shouldn't slap them.

That said, it's going to get very cumbersome to say, "I know not every scientist does this," or "Not every scientist is guilty of that," every time I speak of some failing of science (which is, like, the entire book). I want you to keep in mind that when I speak of scientists or science in the following pages, I'm not trying to paint everybody with the same broad brush. There are some truly exceptional scientists out there trying to rock the status quo, and they should be rewarded and celebrated. Just not in this book, because this book is about sickness.

I desperately want science to *be better*. I want scientists more closely tied to their communities. I want more funding for science. I want more scientists. I want scientists to be a source of knowledge, wisdom, and guidance in our almost incomprehensible world. But I also know that science has a lot of enemies—some imaginary, but some very real. Many people and groups are hostile to science, both its methods and its results. I know that this book, which is explicitly designed to highlight the flaws of science, will be used against the way of knowing that I love. It's an unfortunate side effect of the medicine that I hope this book becomes. But I'm just telling you now: if someone uses my words to sow mistrust and hatred of science, they are abusing both the spirit and content of these pages.

I don't know what your current relationship is with science. You may be an entrenched academic, sipping your morning coffee with Nobel laureates. You may only know of science from the sensational headlines in the Sunday paper. You may be a teacher, or an advocate, or a voracious devourer of all things knowledge. I suspect and hope that, no matter what your background or entry into these pages is, you'll learn something about what science is and how science operates. I know that for some or all of you, not everything in this book will be new to you. But even if you've heard the same story a thousand times, or worse, experienced it yourself, I hope you find some new angle and perspective . . . and that you start advocating for change.

1
The Persistence of Fraud

Publish or perish.

Publish . . . or perish.

Publish (or perish).

Publish, or perish!

Publish or perish ;)

Nope, no matter how you type it, it keeps coming out sinister. "Publish or perish" is the defining phrase of almost every scientist's career. It is the prime mover, the main motivator, the thing that gets you out of bed in the morning . . . and the thing that keeps you up at night.

It's the phrase uttered as a joke even though you can tell from their eyes that they're dead serious. It's the advice given to graduate students when they ask how to succeed in science. It's the excuse you give your spouse when you have to stay late at the lab again. It's the birth of careers and the death of relationships.

It's . . . modern science.

"Publish or perish" is the phrase that most succinctly summarizes what is expected of a scientist in every stage of their career. It means, for the uninitiated, that you need to publish papers, or books, or conference proceedings, or whatever flavor of publication your particular branch of science demands. A lot of them. In "good" journals (and I'll define in a bit what "good" means). It means that if you write a lot of papers, your chances of advancing along the scientific career path are much, much better.

For the academically uninitiated, a "paper" is the primary unit of currency in the scientific world—and also the academic world broadly. It's not a term paper or a final book report that you might have written in high school or college. It's a Fundamental Unit of Science. The purpose of a paper is to succinctly summarize the aims, methods, and results of your original research. What you went looking for in the world around you, how you did it, and what you found. You rarely write one by yourself; usually you write one with a few (sometimes thousand—seriously!) of your closest colleagues. By publishing papers, you are submitting your work to the written record and the scrutiny of your peers.

Why so much emphasis on publishing? Perhaps it started innocently enough with the goal of spreading knowledge and testing theories. But it has since evolved into something else. As we will see, science has turned success into a series of metrics—easily digestible numbers that tell you how well your colleagues are doing and how much harder you need to work.

And if you don't play this game—if you simply don't want to publish for the sake of publishing—you perish. You fail to get into graduate school, or fail to get hired, or fail to get promoted, or fail to win grants. You just . . . fail. You lose the game.

Most scientists just want to be scientists. They want to get jobs and grant money so they can keep doing what they love. To get jobs and grant money and to "make it" as a scientist, you have to publish. On the surface, this kind of pressure is a good thing. It encourages scientists to undertake original research and share their results. The people behind the money get to see what they're paying for, and the sum total of human knowledge increases. Where this system goes off the rails is when the normal pressures to succeed become too intense. If publishing is the primary metric used to judge the success of a scientist, then scientists need to publish at whatever cost. And that creates the opportunity for scientists to gloss over mistakes, ignore alternative threads of inquiry, and even outright lie.

In other words, this atmosphere creates the opportunity for fraud to flourish.

The vast majority of scientists work in university circles. On the surface, "publish or perish" has a simple, powerful motivation: accountability. Universities are shelling out big bucks for all those junior and senior scientists running around. And yes, they do provide some value in the form of maintaining the educational environment, but the senior scientists are what bring in the Real Money: grant funding. Every university takes a slice of every grant dollar. That grant money can add up to millions of dollars—and for some institutions hundreds of millions—every year.

Of course, some of this "overhead" (as it's quaintly called) has a purpose. The university is providing electricity, access, an old chair, pens, and literally a roof over the scientists' heads, and all those things cost money. You are welcome to argue that the amount of overhead taken by a typical university is much more than strictly required for the acquisition of pencils and chalk, and while I won't claim to know the inner financial workings of giant universities, I also won't stand in your way.

No matter how you slice it, money is necessary for the advancement of science.

But some of this has gone off the rails. In some cases, publishers require scientists to *pay to publish research papers*. If you want to publish in, say, the *Astronomical Journal*, then you've got to cough up $110 per page (not to mention the $350 color figure surcharge).[1]

That may not seem like a lot, but a typical paper will run you a couple grand, and you can bet your bottom dollar that your department isn't footing the bill. I've seen this play out firsthand, with my very first research paper published as an undergraduate. My research advisor didn't make tenure and left the university the year I graduated. When our paper was finally accepted, he paid for the publishing charges from his own pocket, even though he wasn't in academia anymore.

The big goal for any scientist is grants. But this leads to two questions: (1) What does it take to get grants? And (2) how do you know if a young scientist will be a grant-winning scientist in the future?

The grant application and award process can be mystifying. A scientist gets a clever new idea to answer some vexing scientific question, like why do stars exist or how to fight cancer. The scientist writes up their plan in a grant proposal. The scientist submits this proposal to Large National Funding Agency. Large National Funding Agency receives the proposal, and a group of scientific peers debates the relative merits of the proposal. They make an award, and the scientist receives some money. The scientist hires students and assistants and buys some laboratory equipment and a new desk. The scientist does science. The scientist publishes science. The world is improved. The process repeats and continues.

Imagine taking that idealized scenario of the grant-to-paper process and looking at it through one of those funhouse mirrors. You know, the ones that make you look all weird and distorted. The actual scientific process contains all those essential elements, but they're . . . well, weird and distorted.

The biggest warping of this process is that, first, before the scientist even get to dream about grants, they need to do a lot of research and write a few papers. If you're wondering how they can do this without grant money, the process starts in graduate school, if not earlier. It's typically in the first couple years of a graduate education that a scientist not only selects their field but the sub-sub-subtopic that will define their research agenda for the rest of their career. Switching tracks midcareer can be a death sentence (because it will take too much time to rebuild your publication game) and is rarely done.

Anyway, resting on the legitimacy of a fat stack of papers, the scientist then approaches funding agencies, but not to propose some radical new idea that will change the face of their field. No, they instead propose research that merely slowly advances whatever it was they were working on already. If they have enough established credibility, they have a good shot at getting the grant money, which allows them to keep pumping out papers, and the cycle continues until the scientist gets tired of spinning the hamster wheel. Since grants are so hard to come by, committees are very risk averse, and papers are a way of signaling that you're good, academically speaking.

In short, in more realistic scenarios, papers are the currency that you can trade in for real money, not the other way around.

What does it take to get grants? A solid publication history.

How do you know if a young scientist will be a grant-winning scientist in the future? A solid publication history.

Universities want to see papers because papers play a big role in determining whom to place bets on—which scientists to continue supporting as they progress in their careers.

Scientists want to best their colleagues, they want to get the grants and the best tenured jobs, and they do that with the only weapon they have: papers.

With all the incentives to publish new research, it's no wonder that the rate of publication has exploded dramatically in the past few decades.

In 2018, there were more than thirty-three thousand active peer-review English-language journals in the world. All told, they published over three million articles in that same year.[2]

That's right. Thirty-three thousand journals, each cranking out almost one hundred articles a year on average, and don't even get me started on academic journals in other languages. I could spend the next two hundred pages of this book simply listing the titles of all the journals in the world, and still not give a comprehensive list. Don't worry, I won't.

Since 1990, the growth rate of journal articles has surpassed the growth rate of scientists in the world. Yes, there are more scientists doing science things every year, but the number of new scientists is decreasing, even flatlining in some disciplines. Meanwhile, the number of new papers continues to climb, meaning that (roughly) the same number of scientists are publishing (roughly) more papers every year.[3]

On the surface, this isn't such a bad thing. More papers = more science = more knowledge = more benefit to humanity.

Except when the papers are useless, or wrong, or fraudulent.

You get what you incentivize: if universities want papers, then they get papers. When I use the word "universities," I don't mean some random building full of academic bureaucrats. I mean the departments themselves, consisting of a typical scientist's friends and colleagues. The same people who serve on grant decision committees, tenure review boards, and student selection panels. "More papers" is exactly what scientists themselves have decided that they want.

The obvious risk of "publish or perish" and the explosive growth of academic publication is that you can get a bunch of junk science.

But before I dig too deeply into actual, real examples of scientific fraud of all stripes and how it pervades all of science, I need to talk about another motivating factor in the scientist's never-ending quest for fame and fortune, which in scientific circles amounts to grant awards and full professorships.

I don't think you're likely to have heard of this little nugget of a phrase: "impact factor." No, it's not a power rating for a cordless drill; it's one of the key metrics scientists use to evaluate the scientific power of journals . . . and their colleagues. We need to talk about impact factor because the story of scientific fraud begins and ends with motivation, and motivation is driven by incentive. We've already seen how the ceaseless hunger for publications creates opportunities for misconduct. Now we're about to see how impact factor ratchets up the whole game.

If you imagined that the underbelly of the academic enterprise would be a seedy world of corruption, intrigue, and tales capable of giving the chills to even the most hardened mafioso, you'd be wrong. Instead, the game of academic politics is played with the scientist's favorite weapon: data.

There are so many scientists publishing so many papers, it's hard to judge what makes a good paper "good," a bad paper "bad," and a mediocre paper "mediocre." And there are so many available journals to choose from that it's hard to judge which journals are the most prestigious. Sure, there are some high- and low-quality outliers that pretty much everyone can agree on. A paper appearing in *Nature* is probably going to have a slightly higher impact on the field than one showing up in, say, the *Tasmanian Journal of Ornithology*. (Yes, I made that journal name up.)

And hence our word of the day: impact.

Every journal in the world, from the most admired and respected to the most are-you-sure-the-publisher-isn't-some-guy-in-a-basement, can have assigned to it a so-called impact factor, defined to be the yearly average number of citations that articles published in the journal get, with that average taken over the past two years. To give you some frame of reference to help understand this arcane formula, if a journal has an impact factor of 1 for a given year, it means that each article published in that journal received on average one citation within the past two years. Complicating matters, the gatekeepers of the index only perform the brain-meltingly tedious calculations for about twelve thousand approved journals,[4] leaving all the other journals languishing without any ranking at all. Clear as mud, right?

The impact factor comes out of the field of (are you ready for this?) "scientometrics," and even though my word processor insists that this is not a real word, it is. Scientometrics is the study of scientific publishing. We're literally at the point in modern-day science where we're writing papers . . . *about writing papers.*

Anyway, the impact factor is used as a handy number—a datum, if you will—to judge the overall importance of a particular journal. The thinking is that journals with high impact factors tend only to accept and publish papers of Great Scientific Importance, and journals with low impact factors will tend to accept and publish whatever documents you can assemble from mashing together your results at the last minute. In between those extremes are vast swathes of important-but-not-noteworthy journals that accept the bulk of publications around the world. So, in addition to the steady stream of papers that every scientist commits to as a part of their progression through life, they really want to publish in "highly rated" journals.

Of course, this doesn't always work out as everyone hopes. Papers appearing in top-flight journals can get zero citations, and papers appearing in no-name journals can become the stuff of legend. We see here the impact factor being perverted from its original purpose—to measure in a crude way the relative importance of a journal—to something driving the never-ending game of one-upmanship. The act of publishing isn't good enough unless you get your paper into a high-impact-factor journal.

This sets up a feedback loop. If a journal has a high impact factor, then you'll only submit to the journal if you think your paper is of Great Scientific Importance. If you think your paper is . . . well, just kinda nice, then you don't go through the rigmarole of trying to publish in one of those high-flying outfits, and instead you send your paper to the editor of whatever you think it most convenient at the time.

The vast majority of my own papers ended up in journals you've probably never heard of, like *Physical Review* and *Monthly Notices of the Royal*

Astronomical Society, because I didn't think that the vast majority of my papers were worthy of a high-impact-factor journal.

There is a huge side effect to this tidy little construct we've assembled to judge the relative importance of journals. It's that we've inadvertently developed a system for measuring not just the impact of journals but the importance of scientists themselves, reducing complex human beings to an individual number. A number that can in certain situations be manipulated by the unscrupulous in order to advance in a scientific career.[5]

You get what you incentivize, and scientists want to publish in high-impact-factor journals because they believe that publishing in those journals will garner their papers a lot of citations, which helps boost their chances of success in their careers. But how do we measure the "success" of a career? Scientists being scientists, they've developed a measurement system to determine how impactful an individual scientist's publications are: the *h*-index.

Basically, the *h*-index is a single number that claims to measure a scientist's productivity and impact. The more papers you publish that a lot of other people cite, the higher your *h*-index. Proponents of reducing complex human beings to a single number claim that the *h*-index correlates with things like winning the Nobel Prize, winning other lesser-known prizes, winning grant awards, and just generally winning at life.[6]

Given that correlation, it's easy to assume that we can safely use the *h*-index in the future to evaluate our fellow scientists. Why, if they've published a lot, and get a lot of citations, then certainly they're on track for the Nobel, right?

Well, correlation is not causation, and past returns are not guarantees of future success. I'm actually quite surprised that scientists, the very people who invented the idea that "correlation is not causation," are using correlations in order to judge and evaluate their peers. If you think I'm making this up, one of the first things a hiring committee will look at is a candidate's *h*-index. If it's too low, or below a magical threshold . . . well, nice knowing you. I have myself heard hiring committee members bemoan the incompleteness and shortcoming of the *h*-index . . . only to use that very metric to judge the quality of a candidate.[7]

Scientists want to win at the "game" of science because they have a lot of other scientists gunning for the same position. To win, they need to rack up a high score and get their *h*-index as high as possible. To get a high *h*-index, they need a lot of citations. Which journals have the highest citation rates? The ones with the high impact factors. To win, they need to publish, a lot, in the best journals they can possibly access.

A lot of people are trying to publish their results in the top-tier journals because the system is explicitly designed to reward that behavior.

And what kind of papers do the top-flight, high-impact-factor journals like to see? The ones that make an impact, of course. The ones announcing new, earth-shattering results. The ones overhauling known methods. The ones that make big claims. The ones that don't just advance science by an inch but by a mile.

Or at least, the flashy ones that people will pay money to read.

This incentive pushes scientists to not bother doing "boring" science, like replication tests or investigating little nuances—you know, doing the hard work of science or simply following their curiosity. Instead, scientists have to push themselves as hard as they can and try to one-up each other like a lifelong game of *Pac-Man*. And they don't just need to write big papers; they need to publish a lot of big papers to get their score as high as possible. This directly leads to rushed, shoddy, poor-quality—but big, important, impactful—papers that appear in the top-tier journals like *Nature* or *Science*.

Scientists are perfectly aware of this. You might think that a paper appearing in *Nature* will be the most rigorously tested, most peer-reviewed, most honorable work that science can produce. But in scientific circles, the saying is, "If it's in *Nature*, it's probably wrong."

Now that we've seen how "publish or perish" and the drive to generate "impactful" results sets up the conditions for fraud to flourish, let's see what that fraud looks like. Scientific fraud can come in many colors, some born of intention and some simply creations of the typical day-to-day incentives of the modern scientific enterprise. I'll call these two kinds of fraud "hard" and "soft," but you can just as well call them "intentional" and "lazy."

Fraud can mean many different things at many different times, and it's much, much more common than you might think, depending on the definition you use. We'll turn to the soft kind of fraud soon, but the hard kind involves outright fabrication or falsification of data with malice aforethought. Remember, data is the golden ore for science. Without data—without evidence—your claims and beliefs are useless. If you want something to be true, you need the data/evidence to back it up. Even if you need to make that happen yourself.

Take, for example, the case of Dr. Marc Hauser, a preeminent professor at Harvard University in the field of evolutionary biology. He studied human evolution, and in particular the roots of conscious thought and whether we share some of those roots with other animals. He was well known, well cited, well liked, and well respected. (I bet he had a pretty high *h*-index.)

But after years of greatness in the field and a few popular books under his belt, exploring topics like the evolution of communication and the origins of morality, suspicions began to arise. Some of the data he was using to make his claims seemed a little . . . off.

Most of the investigation was, and continues to be, under wraps, but some of the allegations have surfaced. For instance, graduate assistants reported to Harvard that Hauser had a penchant for accepting experimental results that he could interpret favorably, while ignoring results that he couldn't interpret favorably. Allegedly, in one experiment, his team would watch as monkeys listened to patterned tones and see if they looked at the speaker when the pattern changed. Leaving aside the discussion of whether monkeys turning their heads is an indication of anything at all, apparently in one case he saw a clear pattern in the monkeys' behavior, while his assistants and students did not. The assistants and students were told to ignore their results and use only his.[8]

Ultimately, in 2010 Harvard University found Dr. Hauser guilty of eight instances of scientific misconduct.[9] Dr. Hauser has not admitted or denied the allegations, but he did resign from his position.

Or take Dr. Carlo Croce of Ohio State University in a case that blew up while I held a position there. Dr. Croce is an expert in genetics and their connections to diseases, especially cancer. By digging through the genetic clues, one hopes, we can get a better handle on treatment to prevent all sorts of horrible maladies and afflictions. It's the usual story with these kinds of high-profile cases: a big-time scientist who has tens of millions in research funds and an army of students and staff and some international recognition—and allegations of faked results.

Those allegations have dogged Dr. Croce's career for several decades, but he (and the university) faced a new set of challenges in 2014 when a virologist at Purdue, Dr. David Sanders, pointed out some interesting features in Dr. Croce & co.'s results.

The issue here comes down to blots. In exploring the expressions of genes, you want to see which proteins are active and running around and which ones aren't. Since the genetic code in DNA determines which proteins get manufactured, you can study the proteins to understand how modifications to the genes can help or hurt a disease treatment. Yes, I'm perfectly aware that this is a highly simplified discussion of this procedure, but please keep in mind I'm an astrophysicist.

Right, the blots. There's this technique called "Western blotting" that sorts proteins by weight and tells you which ones are running around in your sample through the appearance of . . . well . . . blots in a gel.[10]

The thing is, in a paper appearing in a journal, you don't report the blots themselves that you see with your own eyes and that anyone else could easily verify by looking at the same blots with their own eyes, but rather you report pictures of the blots. Pictures that can be manipulated.

According to Dr. Sanders, he found several instances of blots being copied and pasted from here to there, showing the appearance of active proteins where there really was no such thing.[11] That's a big deal because, if true, the core results of Dr. Croce's work would be unfounded, and there would be no reason to sustain his prestige in the scientific community, and the gushing pipeline of grants to his lab.

The grant money is a particularly sticky issue because Ohio State University takes as "overhead" from all his research grants roughly $10 million a year—which is precisely why universities want their researchers to get big grants, and one of the reasons why the urge to "publish or perish" exists. That's not a small chunk of change, even for a giant university. While we don't know the full machinations and rationales within the university, we do know that despite having to issue retractions for a raft of papers, Dr. Croce still maintains his appointment at OSU . . . and his grants . . . and his $804,461 annual salary.[12]

The cases of Dr. Hauser and Dr. Croce are high profile and definitely newsworthy . . . and relatively easy to spot. Emphasis on "relatively." Ultimately, the allegations of fraud in those cases—and countless more that I don't have the time or space to recount, but which most definitely do exist[13]—rest on claims of fake or falsified data, whether either the underlying data simply didn't exist in the first place or were heavily modified to fit a preexisting narrative.

But while this kind of fraud—the obvious, oh-for-sure-it's-totally-fake kind—is well known, it's not the most common kind.

That's why I introduced into this discussion another kind of fraud: "soft" fraud. This is the fraud of the subtle, of the interpretation, of the semantics, of the hundreds of tiny choices one has to make when constructing a research paper, especially one with sights firmly set on increasing one's *h*-index by publishing in high-impact-factor journals. In essence, why bother going to the trouble of cooking the books—outright falsification or fabrication of raw data—when you can simply massage the analysis to give you the result that you wanted all along?

It's a lot harder to sniff out, and it's a lot more common.

Interestingly, this kind of "soft" fraud isn't talked about as much and isn't punished as much as the hard stuff. When scientists discuss fraud, they're being careful to pick a very precise definition of the word, artfully constructed to deliberately exclude the most common forms of academic misconduct (i.e., "soft" fraud) so that they themselves aren't seen to be breaking any moral or ethical code. It's a simple game of arranging the rules such that "fraud" exists, but it's definitely always done by somebody else, and most definitely not done by the majority of scientists, and most certainly definitely not done by themselves.[14]

But "soft" fraud is a lot more common according to whom? I'm not just making things up as I type (which would be a form of hard fraud). This kind of fraud is a lot more common according to scientists themselves *who have admitted doing it.*[15] For example, in a meta-analysis of surveys, when asked, "Do you even make up the data?" (i.e., the all-out "bad" kind of fraud), only about 2 percent of scientists self-report doing something that egregious at least once in their careers. That's already a worrying number. But when asked about soft fraud, like massaging the analysis or monkeying around with citations to fit a certain kind of narrative, that number jumps to 34 percent.

Of course, it takes a lot of courage, self-awareness, and risky honesty to admit that you've committed some sort of fraud, even in an anonymous survey. So when asked about *other* scientists, the numbers jump up to 14 percent for committing hard fraud and 72 percent for soft fraud.

That means almost three-quarters of all scientists are under some form of suspicion from their colleagues of doing something underhanded. Here we see scientists looking at their colleagues, suspicious that they're doing something underhanded to achieve their success, which normalizes the behavior for everyone else to do it. Interestingly, according to this meta-analysis, when the words "fabricate" or "falsify" appear in the survey, scientists are much less willing to divulge their darkest secrets. But when the language is much looser and those kinds of dangerous trigger words are absent, scientists are much more willing to spill the beans.[16]

What exactly do I mean by soft fraud? Well, let me give you an example from my own field, and then I'll talk about how to generalize these cases to other fields. I'm picking an astronomy example here for a couple reasons: (1) this kind of fraud/misconduct/laziness/wishful thinking is very hard to spot unless you're deeply—and I mean *deeply*—embedded in a field and know what to look for; and (2) did you really think you would read an entire book written by an astrophysicist and not get to learn something about astrophysics?

Let's begin with light from the early universe, because that seems as good a starting place as any.

A long time ago (as in 13.77 billion years ago), the entire universe was incredibly small, hot, and dense. It was a pretty packed room: every star, galaxy, and speck of dust in the observable universe was compressed into a volume one-millionth its current size. With no room to breathe, the atoms were ripped apart into a plasma; everything was just a big messy jumble of protons, electrons, and radiation.

Then it stopped. When the universe was 380,000 years old, the cosmos had expanded and cooled enough for atoms to form, releasing a burst of high-energy radiation that persists to the present day—though by now, billions of years later, it's a lot more chill and relaxed. We call this radiation the CMB, for cosmic microwave background. It's called that because (1) it's in space, (2) it's in the microwave band of the electromagnetic spectrum, and (3) it's behind everything else.

The history of the universe prior to the release of the CMB is a bit opaque, but cosmologists are very interested in a particular event known as "inflation," which occurred when the universe was less than a second old and caused it to swell to many orders of magnitude in size—like, from the size of an atom to the size of a baseball. One of the key predictions of the event of inflation is the presence of gravitational waves, which are ripples in the fabric of space-time itself. If inflation really happened, then the young universe should have been flooded with these ripples, which would leave a defining fingerprint pattern in the CMB.

In 2014, a team of researchers using a special telescope at the South Pole—yes, that South Pole—called BICEP (Background Imaging of Cosmic Extragalactic Polarization) claimed the detection of gravitational waves in observations of the CMB, publishing their results in the very high-impact-factor journal *Physical Review Letters*.[17]

It was a big deal. If true, the result would have opened direct observational windows to the first second of the universe's existence, an unparalleled insight. While there were approximately eleventy billion authors on the paper, the leaders of the project were sure to be shortlisted for a Nobel Prize. The news went wild across the nerd world, with article after article exclaiming the importance of the discovery and quotes from eminent scientists gushing about the results.

Like I said, a big deal.

There was just one problem: dust. The presence of random grains of dust scattered throughout the universe (which is more than you might think) *also* makes an imprint in the CMB that just so happens to mimic the signal from the gravitational waves. So, to claim you've seen that precious primordial signal, you have to convince yourself and others that you've cleaned out all the dust first.

The researchers did so with the data available to them at the time, but that data was not the greatest, and they were waiting for results from a much larger survey provided by the Planck satellite mission, of which I was a member at the time. That satellite was making all sorts of maps of the CMB, across the whole sky, but at much lower resolution than the BICEP team could do (in contrast, BICEP was only looking at a narrow sliver of the whole sky, but in higher resolution). A map of the CMB sky as seen by Planck would be critical for assessing the true amount of dust contaminating the BICEP signal.

A couple months after the BICEP team announced their results, the Planck collaboration released their dust maps. And with that, the BICEP-claimed signal disappeared. There was no message from the early universe in their data. Just a bunch of dust.[18]

The BICEP team knew that the Planck results were due out in only a couple months, and they knew that the dust maps provided by Planck would be crucial in interpreting their results, so why did they rush? Why didn't they just cool their heels for a couple months and get it right?

According to one of the lead scientists, UCLA's Dr. Brian Keating, it's because they thought a competing group was about to publish a similar result and wanted to beat them to the punch.[19] After all, the Nobel Prize is limited to only three awardees at a time, and there wasn't enough room on the podium to share it with the competition.

Thus, a rushed job. A bogus result. A big swing . . . and a big miss. A result that everyone in the collaboration knew that they wanted—a detection of primordial gravitational waves—but not enough patience to do it right for fear of getting scooped. After all, the second paper on a topic doesn't get nearly as many citations as the first.

The BICEP gang isn't alone in this. Scientists rush out results all the time, all for a chance to make a splash. This is part of human nature and shouldn't be all that surprising, but the "publish or perish" mindset and the reliance on metrics like the impact factor and *h*-index bring this ugly side of humanity to the fore.

"Now hold on a hot minute here, Paul," I hear you saying as you read this, "that example is all well and good (well, bad, but you get my point), but just a couple bad apples aren't a sign that science is at a real low point and that we have anything to worry about."

Good point, so let me say this: science *is* at a real low point, and we have a lot to worry about.

Fraud in science isn't a new thing. It's as old as science itself and, given that humans will just keep on being human, will outlive whatever science evolves into in the future. Fraud is a part of the human condition and ever present in all endeavors. Science is flawed because it's done by humans. But what I'm talking about here are the incentives that modern science creates, and it's the incentive of "publish or perish" that not only allows for fraud—of both the hard and soft varieties—but actually encourages it.

Soft fraud can take on many, many forms and worms its way into the scientific literature in many insidious, and frustratingly hard to detect, ways. Take, for example, the concept of *p*-hacking, perhaps the most sinister form of scientific fraud. And no, I'm not just making up terms like "*h*-index" and "*p*-hacking" because I like to put random letters before words. These are real things.

Here's how *p*-hacking works. The starting point for science, especially the observational or experimental kind, is data. You collect some data, you analyze the results, you publish (or perish if you don't), you move on.

Some might think, in a naïve view of science, that you only bother to collect the data after you've formed a hypothesis worth testing. Only then do you go out in the real world and decide if that hypothesis has any momentum to it. If not, you chuck it in the waste bin and move on with your life.

That's almost never how it works. Instead, researchers go out collecting absolute tons of data. To mix metaphors, scientists will amass such a mountain of data that they're practically swimming in it. From particle accelerators to sidewalk surveys, I'm willing to bet that there's more raw data out there than all the scientists in the world could ever hope to analyze in their entire lifetimes.

Numbers upon numbers, piled on top of each other. Heaps of scientific data, ready for analysis.

The next thing to typically happen is that a researcher sifts through that data—maybe trying to evaluate a hypothesis, or maybe just looking for something juicy—but they can only report a result if it's significant, in the statistical sense. Using a variety of statistical tools, researchers can claim that a particular result or connection in the data is significant and not just some random fluke (see the case of Venus above, for example). One such tool is called the *p*-value.

If two variables appear to have a connection between them—say, the respondents of a survey who wear yellow shirts on Wednesdays seem to be the same people that eat ham sandwiches on Thursdays—you might wonder if that's just random chance or if there's something deeper going on in peoples' brains. So you break out the *p*-value and do some fancy statistics. The lower the *p*-value, the greater the chance, loosely speaking, that the connection is real and not just a random coincidence in the data.

Typically, a result is only "worth" reporting if the *p*-value is below some arbitrary threshold, usually 0.05.

This is where *p*-hacking comes into play, and it's usually not even a conscious effort.

You amass a mountain of data, with all sorts of different variables in the dataset. Maybe you test a hypothesis—that yellow shirt wearers get a hankering for ham the next day due to some quirk of human psychology—and it doesn't pan out. The *p*-value is too high. It was just a fluke. Bummer.

But . . . there's so much other data that you collected. Maybe there's something interesting going on that had nothing to do with your original hypothesis. So you start looking at different combinations of data to see what gives you a low enough *p*-value that's worth reporting.

Suddenly, you find it! It's wearers of *red* shirts on *Tuesdays* that connects with ham hunger on Thursdays! Eureka! Call up the editors at *Luncheon Meats Quarterly*. This is going to be big.

At first glance, this seems perfectly innocent. You didn't fake any data. You didn't make up any data. It's just . . . right there, you simply had to go looking for it.

Here's the problem. In a large enough dataset with enough variables, there's guaranteed to be at least two variables that are connected with a low enough *p*-value. It's a fact of statistics. You *will* find something that appears to be significantly connected, even if no such connection actually exists. It's just how random data works.

Hence, *p*-hacking. Fishing around in the data for something worth publishing.

In my household, pretty much whenever we encounter a news story that claims that "Researchers discover a connection between *random thing* and *other random thing*," I annoy my family by loudly shouting, "*P*-hacking!" I'm glad they love me enough to put up with me.

This happens all the time.[20] I mean the *p*-hacking part, not the annoying-my-family part.

Let's also not forget sheer laziness. Under intense pressure to publish as much as possible and as quickly as possible, you're tempted not to do the proper checks that you should on your own work. If a result comes out the way you want it to, well then . . . why bother checking to make sure you didn't make a mistake?

If that's not good enough, you have much sneakier tools available to use should you want to construct a particular narrative, like "forgetting" to cite work that contradicts your results, creating an experimental setup that's so unique it's impossible to replicate, getting someone famous to coauthor your paper so nobody bothers you, blaming a junior colleague for being too junior to do good science, and—perhaps the nastiest—simply hiding your results behind a wall of computer code.

Modern science is hard, complex, and built from many layers and many years of hard work. And modern science, almost everywhere, is based on computation. Save for a few (and I mean very few) die-hard theorists who insist on writing things down with pen and paper, there is almost an absolute guarantee that with any paper in any field of science that you could possibly read, a computer was involved in some step of the process.

Whether it's studying bird droppings or the collisions of galaxies, modern-day science owes its very existence—and continued persistence—to the computer. From the laptop sitting on an unkempt desk to a giant machine that fills up a room, "S. Transistor" should be the coauthor on basically all three million journal articles published every year.

The sheer complexity of modern science, and its reliance on customized software, renders one of the frontline defenses against soft and hard fraud useless. That defense is peer review.

The practice of peer review was developed in a different era, when the arguments and analysis that led to a paper's conclusion could be succinctly summarized within the paper itself. Want to know how the author arrived at that conclusion? The derivation would be right there. It was relatively easy to judge the "wrongness" of an article because you could follow the document from beginning to end, from start to finish, and have all the information you needed to evaluate it right there at your fingerprints.

That's now largely impossible with the modern scientific enterprise so reliant on computers.

To makes matters worse, many of the software codes used in science are not publicly available. I'll say this again because it's kind of wild to even contemplate: there are millions of papers published every year that rely on computer software to make the results happen, and that software is not available for other scientists to scrutinize to see if it's legit or not. We simply have to trust it, but the word "trust" is very near the bottom of the scientist's priority list.

Why don't scientists make their code available? It boils down to the same reason that scientists don't do many things that would improve the process of science: there's no incentive. In this case, you don't get any *h*-index points for releasing your code on a website. You only get them for publishing papers.

This infinitely agitates me when I peer-review papers. How am I supposed to judge the correctness of an article if I can't see the entire process? What's the point of searching for fraud when the computer code that's sitting behind the published result can be shaped and molded to give any result you want, and nobody will be the wiser?[21]

I'm not even talking about intentional computer-based fraud here; this is even a problem for detecting basic mistakes. If you make a mistake in a paper, a referee or an editor can spot it. And science is better off for it. If you make a mistake in your code . . . who checks it? As long as the results look correct, you'll go ahead and publish it, and the peer reviewer will go ahead and accept it. And science is worse off for it.

I've encountered this on both ends. One time, after not understanding a result in a published paper, I asked the authors for a copy of their code. They helpfully provided it, and I discovered a mistake in the code. The published results weren't just wrong in a minor sense; the entire point of the paper was invalidated. The published, peer-reviewed article was useless.

My code has been wrong too. In one of my first papers, I discovered a mistake in one of my simulations after my paper had gone to print. Fortunately for my elevated stress levels, it didn't change the conclusions of the paper, but it did lessen the effect I was studying, and I had to submit a correction to the journal. Thankfully, it was a minor mistake, but it was a hair's breadth away from nullifying months' worth of research time.

Science is getting more complex over time and is becoming increasingly reliant on software code to keep the engine going. This makes fraud of both the hard and soft varieties easier to accomplish. From mistakes that you pass over because you're going too fast, to using sophisticated tools that you barely understand but use to get the result that you wanted, to just totally faking it, science is becoming increasingly *wrong*.

What's more, the first line of defense for combating wrongness—peer review—is fundamentally broken, with absolutely no guarantees of filtering out innocent mistakes and guilty fraud. Peer reviewers don't have the time, effort, or inclination to pick over new research with a fine-toothed comb, let alone give it more than a long glance. And even if a referee is digging a little too deeply, whether revealing some fatal flaw or sniffing at your attempt at fraud, you can simply go journal shopping and take your work somewhere more friendly.

With the outer walls of scientific integrity breached through the failures of peer review, the community must turn to a second line of defense: replication.

Replication is the concept that another scientist—preferably a competing scientist with an axe to grind because they would love nothing more than to take down a rival—can repeat the experiment as described in the published journal article. If they get the same result, hooray! Science has progressed. If they get a different result, hooray! Science has progressed. No matter what, it's a win.

Too bad hardly anybody bothers to replicate experiments like this, and when they do, they more often than not find that the original results don't hold up.[22]

This is called the "replication crisis" in science. It first reared its ugly head in the 2010s in the field of psychology, but I don't think any field is immune.[23]

There's a complex swirling septic tank of problems that all contribute to the replication crisis, but the first issue is that replication isn't sexy. You don't get to learn new things about the world around us; you just get to confirm whether someone else learned new things about the world around us. As an added ugly bonus, nonresults often don't even get published. Novelty is seen as a virtue, and if you run an experiment and it doesn't provide a positive result, journals are less likely to be interested in your manuscript. Additionally, because replication isn't seen as sexy, when it is done it isn't read. Replication studies do not get published in high-impact-factor journals, and authors of replication studies do not get as many citations for their work. This means that their *h*-index is lower, which lowers their chances of getting grants and promotions.

Second, replication is becoming increasingly hard. With sophisticated tools, mountains of data, complex statistical analyses, elaborate experimental design, and lots and lots of code, in many cases you would have no hope of attempting to repeat someone else's work, even if you wanted to.

The end result of all this, as I've said before, is that fraud is becoming increasingly common, increasingly hard to detect, and increasingly hard to correct.

The lackluster mirage that is peer review doesn't adequately screen against mistakes, and with the increasing complexity of science (and with it of scientific discourse), fraud just slips on by, with too many layers between the fundamental research and the end product.[24]

It looks like science is getting done, but is it really?

Sure, retractions and corrections appear here and there, but they are far too few and far between given the amount of fraud, both hard and soft, that we know exists.[25] The conditions in every field of science are ripe for fraud to happen, and nobody has developed a decent enough policing system to root it out—in one case, a journal issued a retraction *thirty-five years* after being notified.[26] The kind of investigations that lead to full retractions or corrections don't come often enough, and meanwhile there's just too much science being published. A perusal of the Retraction Watch website reveals instance of fraud after fraud, often coming too late, with little to no major reprucussions.[27]

Nobody has the time to read every important paper in their field. It's a nonstop tidal wave of analysis, math, jargon, plots, and words, words, words. Every scientist writes as much as possible, without really digging into the existing literature, adding to the noise. Heck, take my field, astrophysics, for example. Every single day there are about forty to sixty brand-new astronomy and astrophysics papers birthed into the academic world. *Every single day*, even holidays! According to surveys, scientists claim that they read about five papers a week.[28] In my not-so-humble opinion, that's a lie, or at least a big stretch, because there are different definitions of the word "read," varying from "reading the abstract, the conclusions, and the captions of the prettier pictures" to "absorbing this work into the fabric of my professional consciousness."

The vast majority of science published is not being consumed by the vast majority of its intended audience. Instead, most scientists just skim new titles, looking for something that stands out, or dig up references during a literature search to use as citations. We're just running around publishing for the sake of publishing, increasing our chances of striking gold (having a major paper stand out among the noise and get a lot of citations) or at least building a high *h*-index by brute force of an overwhelming number of publications.

This is exactly the kind of environment where fraud not only exists but abounds. Who cares if you fudge a number here or there—nobody's actually going to read the thing, not even the referee! And the conclusions you draw in your paper will form the basis of yet more work, heck, maybe even an entire investigative line.

So why doesn't it stop? We all know and agree that fraud is bad, that uncorrected mistakes are bad, that impossible-to-replicate results are bad. If we all agree that bad things are bad, why don't we make the bad things go away?

Put simply, there's no reason to do better. The incentive in science right now is "keep publishing, dang it," not to waste your time doing careful peer reviews, making your code public, or replicating results.

I'm not saying that most scientists are attempting to pull off willful fraud. I do believe that most scientists are good people trying to do good things in the world. But with all the incentives aligned the way they are, it's easy to convince yourself that what you're doing isn't really fraud. If you refuse to make your code public for scrutiny, are you committing fraud? Or righteously protecting your research investment? If you had to twist the analysis to get a publishable result, are you committing fraud? Or just finally arriving at the answer you knew you were going to get anyway? If you stop checking your work for mistakes, are you committing fraud? Or are you just . . . finally done, after months of hard work? If you withdraw your article from submission because the referee was getting too critical, are you committing fraud? Or are you protesting an unreasonable peer reviewer?

Scientists aren't really incentivized to change any of these problems. The system is already in place and "works"; it allows winners and losers to be chosen, with the winners receiving grant funding, tenured positions, and fancy cars. Well, maybe not the fancy cars. Universities aren't motivated to change these practices because, so long as their winners continue to bring in grants, it lets them roll around in giant taxpayer-funded money pits (I don't know if this image is totally accurate, but you get the idea).

The very last people who aren't really motivated to change any of this are the publishers themselves. We saw already how most publishers make the scientists pay to get their research printed, and how they don't have to pay for the peer review that makes science all scientific. But the big bucks come from the subscriptions: in order to read the journal, you need to either cough up a few dozen bucks to access a single article or pay prohibitively high fees to get annual access. The only people who can afford this are the big universities and libraries, locking most of our modern scientific progress behind gilt-ivory doors.

Altogether, the scientific and technical publishing industry rakes in about $10 billion a year, often with double-digit profit margins.[29] Of course they don't want this ship to change course. I can't really blame them; they're just playing by the accepted rules of their own game.

There is a way out of this mess, but it won't be particularly easy. The rise of fraud is caused by a host of intertwined incentives, structures, and rewards. But if my thesis is correct—that the pressure to publish at all costs is lowering the quality of our scientific output—then to truly combat fraud we have to target the role of publications in scientific advancement.

Some of the solutions and ideas I'm about to provide are small, simple, and practical. Things any scientist of any rank can begin to put into practice. Some are sky-high ideals that will take years, if not generations, to implement across the entire community. Most are in between. But no matter what, at some level we have to address the "publish or perish" mindset, and we have to do it soon.

To reduce fraud, we need to make science more transparent. With more eyes on more papers, it's harder for bad research to propagate. To make science more transparent, we need to make it more accessible. So, for starters, let's dismantle the stranglehold that the publishing giants have on academic papers. Since papers are seen as the gold standard of academic output, the publishers of those papers get to serve as gatekeepers of "valid" science and the "judge" of quality research. Let this sink in: We've created a system where the ones who get to decide what is worthy science and what is not are the same ones who profit off of the publishing of papers. Thankfully, this is already beginning to change. Slowly, over the course of years, scientists themselves have begun finding ways to circumvent the normal rules of the game. One of those paths has been through the rise of so-called open-access journals, where, instead of hiding behind a paywall, the published research is free and open to all members of the public. Many major publishers now offer open-access journals, but it usually comes with a catch: a hefty price tag for the author, to offset the "lost" revenue of not being able to sell a subscription of the journal to a library. It's not cheap either; the cost to publish an article in an open-access journal usually starts at around $2,000 and goes up to over $10,000.

That's why I'm a little skeptical about open access. It's nice, and I like the spirit of it, but it only solves one end of the publishing and fraud problem by making research papers free to read. It doesn't solve the greater challenge of making academic papers free to create and free to publish. "Open access" can mean a host of different things, and the vast majority of publishers will want to make sure they get their pound of flesh in the end.

So, while open-access journals address the issue of having access to published research, I think we can even take it a step further and insist—and funding agencies have the power to do this—that any paper written under the auspices of a particular grant must be published or made available for free, for all. If we still want to have publishers, fine, but as a compromise a portion of the grant will be automatically assigned to paying the publication charges to ensure that the research is free. This currently does occur in some countries and with some research funding agencies, like the National Institutes of Health, but we need to make it the global and institutional norm.

But even a tectonic shift in the accessibility of science won't eliminate fraud. One, because the incentives to publish bad research are still there; and two, because people are people and bad apples are bad apples. In another prong in the fork of transparency, instead of marginalizing replication, let's celebrate it. Let's put replication studies and repeated efforts on the pedestal, where they belong, as a core practice of the scientific exploration of our universe. Let's begin discouraging the journal practice of "not new enough, sorry." Let's give time in conferences and discussions to replication. Let's—gasp—train our students to engage in replication and duplication and constant cross-checking of the work of their colleagues. Sure, science will be "slower," but again I ask: is that really such a bad thing if we do it right the first time?

I think that the replication crisis may be a blessing in disguise. It awakened a complacent scientific culture that rested on trust, despite the perverse incentives that have emerged over the decades.

Already, many fields are moving to encourage, support, and fund replication. For example, the journal *Psychological Science* now encourages preregistration of studies, to prevent *p*-hacking and fishing for results. The Center for Open Science, founded in 2013, supports researchers in pursuit of replication.

Along with that, let's also celebrate "null" results—research studies that don't end up confirming the hypothesis that they're after. We've somehow forgotten as a community that "no result" is a result, and that "no correlation" is a useful thing for other scientists to know. By insisting that any publishable results be positive, we're practically begging for *p*-hacking and all the other forms of soft fraud that come along with it.

We have the *Journal of Articles in Support of the Null Hypothesis*, which is awesome, but we need to keep pushing for *Nature*, *Science*, and the other biggies to devote a fraction of their pages to these studies.

If I were the Emperor of Science, I would also immediately order all software and computer codes used in the production of science to be publicly available, for free. If you use it to make science happen, then we need to be able to see it, full stop. Yes, I recognize that this might discourage investment in the development of custom software for answering challenging scientific questions—I don't even know how many hours I invested writing my own simulation and analysis software—so I'll offer a compromise: you can submit your software to an online archive like Software Heritage[30] or just slap it on your website, and we, as a community of scientists, will judge you very favorably for it—and will reward you with grants, positions, and promotions, by counting those contributions as positive attributes alongside publishing great papers.

If all this wasn't audacious enough, how about this. I'm going to propose something radical: ending peer review. Peer review of a paper can be a useful thing when it identifies critical mistakes or filters out obviously fraudulent research. But the actual mechanics of peer review are falling far, far short of the high-minded idealizations that initiated the practice long ago.

Let's stop pretending. Let's stop assuming that editors can find qualified referees in a decent amount of time, that referees can provide quality, unbiased reviews, and that it prevents anything that we actually want to prevent.

But how could we have quality journals like *Nature* if we didn't have the practice of peer review? How would they separate the wheat from the chaff?

Well, to tie this all together, why do we need journals?

Ages ago, journals served a powerful purpose: they were the only ones with the wherewithal to print and distribute actual pieces of paper containing scientific information to a wide audience. But this is the modern era, where we have this little thing called the internet—and a wealth of instantly available, globally reached, constantly updated information at our fingertips.

What is the modern journal actually doing?

Providing a filter of quality? Not really. Ultimately, it's the community of scientists that judges whether a paper is good or important or useful, not the publisher's name.

Fulfilling a useful part of the scientific enterprise by putting money in the pockets of scientists? Not really. The peer reviewers are all volunteers anyway.

Copyediting? Not really. Software being software and the march of progress being the march of progress, the vast majority of scientists can now draft up nice-looking papers from the comfort of their homes and offices.

And the peer review part? Well, the *actual* peer review happens in the community, long after a paper is published. Science does attempt to correct itself through time, and colleagues and competitors alike will eventually discover the strengths and weaknesses of papers, regardless of their success at navigating peer review.

You may be surprised to learn this, but many fields of science have pretty much sidestepped journals already. Take a look at arXiv.org, a repository of "preprints," which are drafts of papers that haven't had all the formatting and editing touches that a journal might apply. Every single one of my papers, peer reviewed or not, is available on arXiv. Go ahead. Type in my name in the search box and see what happens. It's all available for free too. No subscription required. Many important papers in astrophysics, for example, have never even bothered with submission to a journal, because arXiv is where every astrophysicist goes to get the latest scoop.

I should say that astrophysicists aren't alone here. ArXiv hosts papers from all fields of physics, from the cosmic to the microscopic, along with math, computer science, statistics, electrical engineering, and more. Outside of that, all medical research papers funded by the National Institutes of Health must appear on the PubMed Central public archive, free and open to the public.

Presumably, scientists want some sort of filter of quality so they're not drowning in a sea of low-quality research. But we are *already* drowning in a sea of low-quality research, because we insist that scientists publish as often as possible. If we move to a community mindset less invested in "publish or perish," there will be less noise to deal with—in the form of fewer, but high-quality, papers—making this process of community interaction so much easier.

Some fields are even taking it a step further by providing platforms for anonymous (if you choose) critiques to be documented right alongside the original papers, like over at PubPeer.com. How beautifully transparent is this? How can fraud or academic misconduct survive—let alone flourish—in this kind of environment?

Those are some practical steps that can be enacted tomorrow if we want to. Even individual scientists can choose to publish only in open-access journals, or not to submit to journals at all. They can refuse to become peer reviewers for a journal and just send comments directly to authors. They can join online discussions. They can insist that hiring committees change their practices. They can educate their students on the importance of scientific integrity and replication—and serve as role models themselves.

But we can take it one step further. To truly eliminate fraud, we need to critically examine how scientists approach advancement, recognition, and rewards. We need to change our culture for the better.

For example, let's all stop getting so excited about impact factors and the *h*-index. In fact, let's just take the entire concept of the *h*-index and toss it out the window. The point of the *h*-index was to give scientists an easy way to measure the "importance" of their colleagues, which was a fool's game to begin with. How can you possibly hope to reduce something as complex as a human being's scientific influence with a single number, no matter how contrived and elaborate? You just can't, and the rise in popularity of measures like the *h*-index has created a toxic atmosphere.

So let's stop pretending that the *h*-index provides useful, bias-free information—because, surprise, the *h*-index is favorable to the scientific elites who are already in charge—and find other, more organic ways of judging candidates for grant money and positions. Maybe we could, I don't know, actually talk to them, and talk to their peers, and make our lives just a little bit harder in exchange for making the world a little more just.

That's right, we could interview more candidates. The way committees usually structure themselves is to receive a few hundred applications for a single open position. They then use various metrics to whittle down the candidate pool and finally interview the cream of the crop. One of the reasons that this process requires so much whittling is that there are way too many junior scientists applying for positions, and way too many scientists overall competing for advancements and rewards. That's a separate issue that I'll tackle later. But for now, we can invest more time and energy into discovering who these candidates really are, as people and well-rounded scientists, rather than the one-dimensional number of the *h*-index.

And if any department chair or grant committee panelist wants to argue that they don't pay that much attention to the *h*-index or publication count and that they already take a more holistic approach to candidate selection, then I challenge them to post their decision-making process on their websites, plain and clear for all to see. The fact that *literally nobody has done this already* should raise alarm bells.

And while we're at it, let's burn "publish or perish" to ashes. The only reason we devised concepts like the impact factor and the *h*-index was because we believed that the only way a scientist could demonstrate their worth was through the production of papers, that if they didn't keep producing papers at a continuous pace throughout their careers, they somehow stopped being insightful, respectable, industrious, worthy scientists. "Publish or perish" would have strangled the careers of many scientists in the past—scientists whose work revolutionized their fields and our understanding of the universe.

We could, in a single moment, come together as a community of scientists and decide, altogether, that "publish or perish" is officially dead. We haven't, due to all the disincentives that we've explored here, but that doesn't mean that we can't. It wouldn't be easy; there's a strong element of bureaucratic entrenchment motivating things like *h*-indices and impact factors. With so many scientists writing so many papers, it's easy to fall into the trap of metrics designed to summarize, rank, and reward.

All it would take is a single conscious effort to stop emphasizing publication counts and citation scores, and most of the problems associated with fraud and misconduct would simply vanish into the margins of history, a decades-long "oopsie" on the part of the scientific enterprise. Sure, we'd get "less" science, as measured by the number of papers published every year, but would science itself worsen after a decision like this? I doubt it.

Most of this mentality shift will have to come from new scientists, simply because established, senior researchers have little incentive to change a system that they are currently succeeding in. So what can new scientists do? Individually, not much, because as long as the structures remain in place, any choice they make that goes against the grain will make them appear less viable as candidates to the ones who are controlling the game. So maybe the best choice for junior scientists is to not play the game at all. Leave. Quit. Go do something else.

In the end, the "publish or perish" mentality cheapens science. Let's make it worth something again.

As we've seen, some change is already happening, like the San Francisco Declaration on Research Assessment,[31] which recommends against journal rankings, and the Hong Kong Principles,[32] which encourages member institutions to use more than publication count to measure success. But these efforts are nascent (at the time of this writing, only one American university has signed the San Francisco Declaration); they still need a lot of time, intention, awareness, and room to grow. Endorsement by major institutions and funding agencies would be nice too.

Fraud will always be with us. It's part of human nature. But we can work to eliminate soft fraud, which is caused by the "publish or perish" mindset. That way, (ideally) the only fraud that remains is the hard kind, which can be quickly identified and punished via open and free access to scientific results and methods, and a culture that celebrates and rewards replication and null results.

Fraud will always be with us, but it doesn't have to be *encouraged*.

2

The Avoidance of Risk

It was in Paris that I learned to love the void.

NO, IT WASN'T AN EXISTENTIALIST CRISIS—it was a curiosity for a new line of research. I had just arrived there in 2011, with a freshly minted PhD in hand, to start my postdoctoral appointment at the Paris Institute for Astrophysics with Ben Wandelt, who would be my postdoc advisor for the next three years.

Ben and I had two venues for extended conversations: his office, and out for lunch. And so one afternoon over a second cup of espresso he told me about this interesting little side project he had been working on with another postdoc, Guilhem Lavaux. They had discovered a fascinating little property regarding a certain kind of structure known as cosmic voids that could make them useful for understanding the larger universe.

Cosmic voids are one of the largest *things* found in nature: vast expanses of just about nothing stretching twenty million light-years across and more. By far, the majority of the volume of our universe is void. All the galaxies pile together in filaments, walls, and dense clusters, taking up relatively little space in a pattern known as the cosmic web.

The cosmic web is super cool and super pretty, and also intimately tied to the history, ingredients, and future of the universe, and so astronomers had been performing surveys of the cosmic web for decades. But while the voids had been known about for that same amount of time, they had been largely ignored. Yes, we knew they existed, but cosmologists were much more interested in galaxy positions and properties.

But Ben and Guilhem had found something curious about the voids: their shapes were surprisingly sensitive to the ingredients and history of the cosmos. It turned out that the more you studied nothing, the more you learned about something (a phrase I would use to cap off my talks for years to come).

I was hooked. I started learning about voids, about structures, about algorithms, about surveys, digging into a side of cosmology that I had largely ignored for my PhD research.

Ben hired more postdocs and graduate students, and we worked. We developed techniques for finding voids, we published catalogs of voids so the community could share in the work,[1] we investigated every possible corner of void properties, turning them from vaguely interesting cosmological side dishes to a powerful tool for the next generation of scientists. I traveled like crazy, proselytizing on the value of voids and building networks of collaborators.

I was making a name for myself as "the void guy," and I thought this was putting me in a good position to be competitive for faculty positions. After all, I was publishing, not perishing. I was meeting lots of people and building a network of collaborators and coauthors. I was contributing to the field in a not insignificant way. I was leading and advising graduate students. It was interesting work. It was new work. We could feel, day by day, a sense of collaboration growing and a wider scientific community growing in interest.

I was also advised, by multiple senior researchers, to knock it off.

I'm not joking. I was told clearly and explicitly, by people I trusted, that I should pivot away from voids if I wanted to make for a decently competitive faculty candidate. I was told that I was too busy chasing after objects of my own curiosity, and I wasn't really serving the greater scientific good. There was real work to be done in many important, relevant, community-approved directions, and voids weren't one of them. Yes, I was becoming "the void guy," but nobody wanted to hire "the void guy" as a professor.

I'll try to be as impartial as possible for the rest of this chapter, but you'll have to excuse me if things get a little heated, as I believe I received some noxiously bad advice from otherwise well-intentioned people, because those people were caught up in a modern scientific system that avoids risk as much as possible. Many—if not all—modern fields of science have become extremely risk averse, almost to the point of absurdity, moving forward in new directions not with energy and agility but with all the lethargy of an entrenched and encrusted bureaucracy.

The idea that science is risk averse at first seems like a contradiction. How could the field of inquiry that, by definition, is designed to challenge our long-held beliefs and let our curiosities wander as far and wide as possible in the name of knowledge be cautious? Science *as a philosophy* is all about exploration, so how could science *as a field* want to explore as little as possible?

Note that almost nobody defines—let alone discusses—"risk" when it comes to scientific pursuits, so for this chapter I'm going to use a lot of stories and examples to demonstrate how risk manifests within the scientific community.

In my personal case, I was seen as a risk because I was young and unproven. This is the first manifestation of risk in science: the simple bias against the young.

About the PhD part: yes, it's true, a PhD is not enough, as evidenced by, say, books literally titled *A PhD Is Not Enough!*[2] Most fields of science nowadays require several years of term-limited appointments to (1) minimize risk and prove that you'll be a good enough candidate for a tenured or permanent positions and (2) filter out as many people as possible, a topic we will explore to our dismay in the next chapter.

Hiring committees and grant award panels—by which I mean *other scientists*—want to make safe bets, not risky ones.[3] They would rather know that a candidate or research program will make small but certain progress rather than large but uncertain advancement. You simply don't know if a young researcher will pan out and provide a solid scientific return on investment. Permanent positions aren't cheap and are very rare—we need to make sure they go to the Right People. And so young scientists carry more perceived risk.

And I haven't even brought up scientists from poorer countries, which face almost insurmountable odds in getting their ideas out into the wider scientific community. Scientists in developing or marginalized countries don't just have to fight the usual battles that all scientists face, but they have a harder time joining large collaborations (because they weren't involved in the initial stages and can't just buy their way in), they have fewer opportunities to bring visibility to their work (even if they could publish in an English-language journal, basically nobody reads articles anyway), and—because of the circumstances of their home country or institution—are seen as *risks*. They're not safe bets because their university might lose funding or their country might become unstable. In a foreshadowing of what we'll see in later chapters about the lack of diversity in science, they get caught in a Sisyphean struggle: since they're never seen as safe bets, they *become* risky through no fault of their own.

The second problem, and the second manifestation of risk aversion, was that void research was nascent and in the margins. Serious cosmologists focused on galaxy surveys, not void surveys, and were part of big collaborations, not small networks of independent researchers. Sure, I was gaining a name, but not in the right circles. We were making a case for voids—but a case is not a conviction, and a career defined by cosmic voids wasn't enough to make a bet on, even for grants, let alone a tenure-track faculty position.

As fields of science grow, they tend toward larger collaborations, and those larger collaborations are safer (in terms of risk) . . . and much less transformative.

At the same time I was studying big patches of near absolute nothing in the universe, I was also working with a giant group of collaborators centered around the Planck satellite mission.

In contrast to my work with voids, which involved small networks of coauthors, the Planck collaboration was . . . huge. Hundreds of scientists from around the world had spent nearly a decade to design and launch a satellite, named Planck in honor of physicist Max Planck. The satellite itself, led by the European Space Agency with help from NASA, was orbiting far from the Earth, observing an ancient light left over from the earliest moments of the universe—the same cosmic microwave background, or CMB, that we met in the previous chapter.[4]

The purpose of Planck wasn't to boldly go forth into uncharted territory and uncover new mysteries of the universe. No, it was designed from the beginning to measure things *that had already been measured*, but with more precision and accuracy than had been achieved previously. Yes, this is still science, but it's a gradual, inching form of science, not a rapid, agile form of discovery. Nobody expected anything massively new or noteworthy from the Planck mission, unless we came upon something by accident, except for an updated list of measurements of various properties of the CMB and the universe.

This is exactly what happens to mature, developed fields of science. There is less and less interest in taking start-up risks, and more of an interest in creating . . . well, for all intents and purposes, a science-themed corporation. Planck was not the first (and if current proposals have their way, not the last) orbiting mission to study the entire CMB sky. It succeeded an early 2000s mission called WMAP, which itself succeeded an early 1990s mission called COBE (the leaders of which, but not the dozens of junior researchers, won a Nobel Prize for their work).

Within the Planck collaboration, I was a cog in a vast machine: responsible for one component of the "pipeline," the path from raw instrument data to useful science analysis and output. I played the role, along with dozens of other postdocs, of low-level software engineer, hacking out code in support of a mission that was far too large for any single research group to command.

In the end, Planck did not yield all that much new information. Sure, we knew some numbers better than we knew before, but no new bold insights had revealed themselves in the analysis. Even the press release for the first round of results was decidedly mild, using the word "confirm" far more than the word "discover."[5]

The Planck mission is a prime example of what I call *incremental* science. The Planck collaboration fully knew their instrument would likely not deliver giant leaps in knowledge but rather a refinement on what we already knew. But that's precisely what made it successfully win the funding to make it happen: it was a safe bet. It was low risk. With Planck, you pretty much knew what you were going to get before it even left the launchpad; all that was left was to competently perform the analysis, and there was plenty of money available to hire grad students and postdocs to do just that. Everyone could write a slew of papers, and the universe—and the academy—could go on doing its thing.

Was Planck worth the seven-hundred-million-euro price tag? It depends on what you mean by "worth." We certainly learned more about the universe than we did prior to the mission, but I ask you this: what else could we have learned with seven hundred million euros?

Lastly, I was a risk because I was studying something most other people in my field were not studying. It's not that void work is exceedingly arcane or based on high-level mathematics; it just wasn't *popular*, for no reason in particular. And if a topic isn't popular, the group consensus is that it's risky. This is the last manifestation of risk aversion: simple human inertia.

Science is usually incremental—it's much easier to slightly push on a boundary of knowledge than to go off completely into the wilderness—but modern science is now largely incremental *on purpose*.

Ultimately, the ones who judge and support scientists are other scientists. Scientists make up hiring committees. Scientists sit on review panels. Scientists award degrees and confer honors. And the scientists who sit on those committees and on review panels are the ones who "won"—who followed a particular track through the maze of scientific academia, and while their own work may require leaps of imagination and dizzying feats of curiosity, they—as a group—move as slowly and carefully as humanly possible.

Newfound ideas are risky. Established lines of research, with broad networks of collaborations and gallons of institutional money poured in, are safe. We know what we're going to do, we know how we're going to do it, we know who's going to do it (typically, senior researchers providing guidance and junior researchers doing the grunt work), and we—oftentimes—already know what the results are going to be.

But new ideas? New lines of research? New techniques? Sure, those *might* pay off . . . someday, if they're both good and lucky. New ideas face an enormous institutional bias against them: in order for new research directions to gain in popularity, you have to convince other researchers to stop or slow down work on proven directions and take a small personal risk themselves—which is no small feat.[6] So in order for a new technique or idea to become popular, you have to invest enormous amounts of personal time and effort (and money) into it—and it may not work, in the end.

All this amounts to a surprising amount of risk aversion in the modern scientific enterprise. This is a risk aversion born out of the desire not to rock the boat, not to spend grant money on potentially unfruitful research lines or people. This avoidance of risk leads directly to stagnation and, even worse, decades of research wasted in certain directions because "that's what everybody else is doing." Constant risk aversion drives good scientists away from the field, drives good science away from our knowledge, and drives good public awareness away from science. And it doesn't have to be this way.

To demonstrate just how bureaucratic the scientific process has become in the modern era, I submit for your examination Exhibit A: The Decadal Surveys. Since the mid-twentieth century, every ten years the scientific worlds in anything astro-related come together to decide their priorities for the next decade, in the form of a report (because, of course, how else could such a rigid process deliver its recommendations?).

The whole process is organized by the National Research Council of the United States National Academies and is submitted to such august organizations as NASA and the National Science Foundation to help them set funding priorities and guidelines for the next generation of scientists.[7]

On the surface, the instrument of the Decadal Survey seems at worst benign and probably beneficial. There are a lot of astronomers and physicists running around, doing their research thing, with little to no overall coordination or planning. The funding agencies (aka the people paying the bills) want to know what everyone's been up to and what they plan on doing in the next few years, to make sure it lines up with funding levels and to help those agencies make their case to Congress for more moolah.

Okay, so far, so good. An actual honest-to-goodness survey, describing what people have been up to for the past few years and what they'd like to keep working on, sounds fantastic, especially since modern science is fueled by giant bureaucratic government agencies. And what giant bureaucratic government agencies want, giant bureaucratic government agencies get.

But wait, hold on. A survey of the current status of research is one thing, and probably a good thing, but a breakdown of desires and plans for the next *ten years* is a completely different beast, and this is where all the risk is taken out of modern astronomy.

The survey works like this. Scientists from across the United States nominate and select the committee to run the thing—and lo and behold, those committee members are largely well-known, well-respected, well-funded, deeply entrenched experts, chosen by other well-known, well-respected, well-funded, deeply entrenched experts. After all, in the meritocracy that science pretends to be (more on that later), if you're successful at research, then that makes you a "leader," regardless of your other qualities or attributes, and so perfectly suited to fashion the report that guides the next ten years of science.

Then subcommittees, panels, and other various organizational paraphernalia pop up, to discuss whatever Hot Topics astronomers are getting excited about at the moment. Those subunits fashion subreports and also put out calls for individual scientists (or collaborations) to submit white papers to make cases for (or against) certain research lines.

After an inscrutable and opaque process, a "survey" pops out of these machinations and gets circulated through all sorts of . . . well, inscrutable and opaque government institutions. Now, on the surface, everything about the survey claims that this explicitly does *not* guarantee funding and that priorities can change in a moment's notice. In reality, if your project or idea is identified as a priority by the survey, you're going to get priority for funding.

What this ends up being is an exercise in scientists convincing other scientists that *their* research priorities should be *community* research priorities, like a giant round of prefunding proposals.

By forcing a community consensus to get prioritization in the survey, the Decadal Survey demands slow, incremental work based on already established research lines—because that's what the people involved in the survey are already working on, and what they want to continue working on. The entire program is a pat on the back from astronomers to themselves, congratulating each other on a successful ten years and raising a glass to another decade of fruitful endeavors . . . that they were going to work on anyway. Despite the frenzied activity surrounding the preparation of the survey, it's a deliberate avoiding of any question or problem that the powers that be might deem inappropriate.

The Decadal Survey process is explicitly designed to marginalize unpopular or new research lines or ideas in favor of entrenched consensus directions because it's run and organized by people performing entrenched consensus research. And entrenched consensus research only grows in new directions slowly, carefully, deliberately, favoring incremental progress over bold new ideas (like we saw with the Planck mission). And, as we will see, funding agencies *love that stuff.*

Just in case you were wondering, astronomy isn't alone in constructing these monstrosities. Decadal surveys inform grant-making decisions in materials science, essentially all fields and branches of physics (including biophysics, high-energy physics, and nuclear physics), Earth science and climate systems, biological and life sciences, social and behavioral sciences, cognition and perception, and more.

But this kind of groupthink happens even without an overbearing steering committee (even the concept of a "steering committee" seems anathema to the core values of science), as I now present to you as Exhibit B from the land of astronomy and physics: supersymmetry.

Throughout the twentieth century, physicists uncovered the fundamental forces of nature and an absolute zoo of fundamental particles. These forces and particles combined make up the entirety of our physical existence, which is kind of cool. But despite decades of work, some mysteries stand out. For example, we are able to describe three of the four forces of nature through the language of quantum mechanics, but we can't do the same for the force of gravity. Why? Also, there are two large "families" of particles that don't have any relation to each other whatsoever. Why? And so on—modern physics is riddled with these kinds of unsolved problems.

Beginning in the 1970s, an idea called supersymmetry developed, which linked the two great "families" of particles together through some deep (and complicated) mathematics.

Over time, physicists came to a certain consensus: supersymmetry was basically a foregone conclusion (*of course* nature will be this elegant on a fundamental level), we just need a particle accelerator big enough to detect it and we'll be all done.

The physics culture was so focused on supersymmetry that competing ideas were pushed to the sides, *despite* a lack of experimental evidence one way or the other, which should have been the ultimate arbiter of the debate. Instead, supersymmetry was *in*, and if you wanted to advance your career as an aspiring high-energy theorist, you followed the trends. Spending time on other avenues was too risky.

Without anybody making a grand, committee-level decision, the entire field of high-energy physics spent decades working out the math of supersymmetry. If you were a young, promising theorist, your options for education and training were limited since any advisor you could go to in graduate school was likely working on supersymmetry. And all the conferences were about supersymmetry, so your options for networking and collaborating were limited. And all the funding agencies were paying for supersymmetry research (because that's what they had been convinced to fund), so your options for cash were limited.

So you became a specialist in supersymmetry, pushing on some tiny little corner of the ideas rather than making a path of your own.

Eventually, we took $4.75 billion and a decade to build the Large Hadron Collider at CERN, on the Swiss-French border, and smashed tiny little particles together at nearly the speed of light to see what happens, searching for signs of supersymmetry (among other things).

That search for supersymmetry began on September 10, 2008. As of the writing of this book, over twelve years later, there are no signs of supersymmetry.[8] And not just "no signs so we better keep on looking"; I mean that supersymmetry is largely *ruled out*.

Supersymmetry may be the greatest failed theory in the history of science. Forget the nearly $5 billion it took to open and run the Large Hadron Collider. Forget the decades of serious work and sweat that theorists poured into the ideas. Focus on what *didn't* happen because it wasn't fashionable. How many ideas were snuffed out, extinguished, ignored because supersymmetry was the safe (and, ultimately, wrong) bet? What *don't* we know about the universe today that we could've known if theoretical physicists had been willing to take more chances?

We will never know.

Let me share one more example before you think that I'm writing this entire chapter because I'm bitter about the advice I was given about void research. To give you some perspective on the risk aversion in modern science that has real-world, life-or-death consequences, let's take a look at the current state of cancer research.

We've spent billions upon billions of dollars as a society trying to fight cancer, and yes, we've had some notable successes. Deaths from breast cancer—one of the most common forms of cancer—have dropped by 35 percent since 1975. Colon cancer too has been greatly subdued, with incidents (and deaths) dropping 42 percent since 1985. You are now 35 percent less likely to die of prostate cancer than you were in 1975.[9]

This is fantastic! "Cancer" is much less of a bogeyman that it was just a generation ago, and we have the tireless efforts of scientists, clinicians, practitioners, and even patients to thank for it.

But (and you knew it was going to be a "but") most of that decline in cancer death rates can be attributed to better prevention and more aggressive screening, nipping cancers in the bud before they get too destructive. To be fair, that knowledge took a lot of science to uncover and solidify, so I'm not going to take it for granted. But cancer is still the second leading cause of death in the United States, claiming almost 600,000 souls in 2018.[10] And many forms of cancer, like melanoma and non-Hodgkin's lymphoma, are just as deadly today as they were decades ago, despite truckloads of money spent trying to fight them.

It's worthwhile to ask the same important questions that we ask of any research in physics or astronomy or anywhere else: Have we spent our money wisely, and could we have done better? Are the multiple overlapping hierarchies of steering committees that guide national-level research doing the best job they can?

Again, it's impossible to say, but many cancer researchers are raising serious questions about the scientific practice that they find themselves in and wondering if structural problems (like risk aversion) might be to blame for the lackluster progress we've made in the past few decades.[11]

Stories like these from across every field of science could fill this entire book, but we'll just have to leave them to references.[12]

What causes the gears of scientific progress to slowly grind to a halt? Philosophers and historians have spent several decades (if not centuries) trying to understand scientific progress. As we know, science itself is a relatively new way of thinking and organizing information about the world around us, and by all accounts has seen remarkable success in its short life. In four hundred years we've gone from a small, Earth-centered universe to a cosmos 13.8 billion years old and 95 billion light-years wide. In four hundred years we're gone from the four humors to the germ theory of disease and the vaccine. In four hundred years we've gone from "animals are just, like, *there*" to the theory of evolution and dinosaurs.

I would say, and I'm not just recommending this because I'm a scientist myself, that science deserves a pat on the back.

Given the remarkable nature of the expansion of scientific knowledge, thinkers have pondered exactly how we (as in, scientists) go about it. How do individual scientists choose research problems? How are results disseminated and accepted? How do paradigms shift and new structures replace old ones?

You won't be surprised to learn a couple general themes. One, young scientists do not come up with new research ideas in a vacuum—they are influenced and guided by both senior researchers and the general milieu that they find themselves in (i.e., whatever topic is fashionable).[13] Second, when a new field opens up, the easiest problems (the "low-hanging fruit") get solved first because they're easier, leaving the harder problems for later generations to tinker over.[14]

Science is not about individuals sitting alone in a vacuum-chambered laboratory, coming up with radical new ideas, testing them, and proclaiming them to the world. At some points in the history of science, you may be able to identify characters that could kind of, sort of, fit that description (if you squinted hard enough), but that is simply not how it is. Science is about networking, learning, growing, collaborating, discussing/arguing, and communication.

When a new field—or branch of a field—emerges, it's worked on by relatively few people working in relative isolation, making for small, isolated collaborative networks. A trio of researchers in this laboratory over here, a duo at that university over there. Maybe they don't even know each other exists. When I first started working on voids, that's exactly what I discovered: little pockets of void-related research occurring all over the globe, without anybody really talking to each other. Those isolated groups tend to work on the problems that only small, isolated groups can solve: the low-hanging fruits. This partly explains why our void group in Paris was so productive. Nobody else had dug into voids the way we did, so we could run around writing paper after paper, exploring to our heart's content.

But as a field evolves, two things happen. One, the network of collaborators grows, linking more researchers at more institutions. This is good—more people means greater capacity to solve harder problems—but it also increases the "deadweight," the amount of time spent networking and communicating rather than discovering and experimenting.[15]

Think of it like this. If you're making a bed by yourself, it takes a certain amount of time. If you have a helper, you can get it done roughly twice as fast. But if you have twenty helpers, you spend more time communicating and coordinating than actually getting the dang bed made.

What's more, since the low-hanging-fruity problems have been solved and written about, that leaves only the harder problems. The ones that take more time, energy, and money to tackle.

Young fields have more "edges"—there are fewer people with more opportunities to tackle novel problems. As a field matures, those people blob (for lack of a better term) together, increasing the amount of collaborative networking but reducing the number of "edges" where people can push in new directions.

Just like people, fields of science slow down in their old age.

And just like people, fields of science become more averse to risk in their old age.

As fields become more *mature* (to put it politely), they grow larger and the members of that field start to work on harder problems. New recruits into that field are trained on the results of their glorious ancestors and set to work making minor additions to the knowledge base that defines that field. Those larger problems require more financial resources to solve (bigger laboratories, bigger telescopes, bigger atom-smashers, etc.), and only large, national-level funding agencies can provide those resources. Those funding agencies don't want to make multimillion-dollar bets if they're risky, so they insist on safer choices (we'll get to exactly why they insist on that in a bit).

Simultaneously, those fresh new researchers into the field have to work on smaller and smaller edges of the problem domain specific to the field, becoming so much grist for the mill. Mature fields of science are impossible to do solo. Young researchers aren't really taught to be independent, free-thinking explorers of the universe; they're part of a *team*. Solo-author papers are the exception rather than the rule in modern science. Young scientists also become super-duper-hyper-focused on tiny subproblems of the larger picture. This is because, one, that's all that's left for them to work on. And, two, young scientists are trying to protect their careers, making a name for themselves without making waves. The only way to do that is to focus on problems that aren't seen as too risky—by completing hyperfine subtasks assigned to them by their superiors in a large collaborative network.[16]

Want examples? The cosmic microwave background was detected by a couple of radio engineers *by accident*. Fast-forward half a century, and the Planck mission cost about a billion bucks and employed hundreds of (temporary) young scientists. The structure of DNA was discovered in a run-of-the-mill laboratory setting. I can't even type the number of zeros required to express how much we spend on genomics research nowadays. And so on.

Modern science is almost—and perhaps already—too complex. In so many fields, from biology to cosmology, no single human has the mental capacity to encompass all the problems that the field is trying to solve. When the low-hanging fruits are all gobbled up, that only leaves the difficult problems. This makes modern science lethargic and stagnant—and as studiously risk averse as possible, because of a general sentiment of *we've come too far already.*[17] Nobody wants to take a risk on a new field or branch of research, because (1) it might not pan out and (2) we've already sunk millions of dollars and entire careers into the existing, *proven* field. Better to keep going slowly than take a risk at not going at all.

It's no wonder that computers have taken such a central role in the modern scientific undertaking—we're literally studying problems that are too difficult for human brains to solve. The modern scientist is really an amateur software engineer, something that I'll talk more about later.

But the simple evolution and maturation of scientific fields isn't the only reason that we're so risk averse in modern times. No, we're also here in this moribund state because we, collectively, decided that we should be. And we did it through the most powerful force of all: love.

I mean, money.

It is, unfortunately, the powerful incentivizing force of *filthy lucre* that cements mature fields in their place, forcing them to understand the universe as slowly as possible—and choking off any potential contenders.

Let's say you're a fresh-faced scientist in a brand-new academic faculty position. You're ready to do some Serious Research. You get a salary—your university is going to pay for your lavish lifestyle for the nine months out of the year that you show up to teach classes. If your institution is wealthy enough, you'll likely start off your rewarding scientific career with a start-up grant, also provided by your university, to accomplish things like pay for research time in the summers, hire a student or two, buy a pencil sharpener, and so on. But once those funds run out (and they will, faster than you might realize), if you want to do things like pay for research time in the summers, hire a student or two, or buy a pencil sharpener, you need a grant.

In the United States (and I'll focus the rest of this chapter on the United States) you can apply for grants from the US government, your home institution, and private foundations. The private foundations may be rich, but they dole out peanuts compared to institutions (averaged over all of science; there are always hyperspecific exceptions like when the Gates Foundation decided to try to cure malaria worldwide). Universities have been spending more and more over the past few years, and have been largely responsible for the increase in research funding nationwide. But that initially rosy picture carries one major caveat: the vast majority of that new funding is going to biomedical science (e.g., pharmaceuticals) with the explicit goal of developing patentable (and profitable) new drugs.[18] In this sense, the universities are acting more like corporations than like institutions of open-ended research, so I'll leave that to the side as I believe that the troublesome issues that arise from that approach speak for themselves.

If you want to engage in fundamental research, you have to ask the behemoths that are the National Science Foundation (NSF), the National Institutes of Health (NIH), the Department of Energy or Department of Education (DOE and DoE, respectively), and NASA.

You then look for open *calls*, or requests for proposals, under various programs put on by those agencies. Some of them might be rather open-ended, accepting any qualified applicant for any bright idea that pops into their head, and some might be rather closed, with proposers guided to a specific research goal prioritized by the Big Brains upstairs.

The very existence of the second group of options for proposals—specific requests with a known goal in mind—needs no further discussion, as it's already obviously risk averse: government bureaucrats deciding the research agendas before the research has even happened. Not that I have anything against government bureaucrats, mind you (my dad was one), but I don't think that they are going to be best positioned to maximize the potential for exploring brand-new branches of science.

But even the more open calls for proposals are pretty straitlaced. The audience for your proposal—where you make a case for the scientific merit and benefit of your project, the amount of resources that you'll need, and the argument that you're the best person for the job—is your peers.

By and large, the people reading and judging your proposal will be other scientists in your field, and most likely *senior* scientists in your field. According to the NSF, only 19 percent of proposal reviewers in 2018 had the distinct pleasure of this being their first time reviewing, indicating that the vast majority of people scrutinizing your proposal will be . . . *seasoned.*[19]

Did I mention that these senior scientists will have their own ideas of what constitutes "good" research and will also be competing for those same precious grant dollars (not in the same year as that would be a little too on the nose for conflict of interest, but the point still stands)? So we have a situation where senior researchers are deciding the fate of junior researchers, walking into the whole situation with preconceived notions of what kinds of science they would prefer to see.

Oh, yeah, they're all usually volunteers too, so this uncompensated work has to compete with all the other tasks demanded by their regular jobs.

Like I said earlier, nobody in science really talks about risk in any meaningful way. There are no high-level discussions evaluating the pros and cons of "risky" research, or what "risky" even means. The lack of discussion is striking—it's hard to find a profession more interested in itself than science (and I'm guilty of that too, by very nature of the book you're reading right now). But as far as I can tell, nobody at the funding level has done any formal surveys or studies examining risk (and aversion to risk) as a motivating factor in how grant dollars should be allocated, and there hasn't been any community-wide conversation about this subject. In other words, nobody has out-and-out stated that "risk" in science is frowned upon.

Instead, we have to look at the clues and see what funding agencies (and the reviewers who make the recommendations) *actually do*, regardless of what they *say* they do.

For example, in the NSF's own "Merit Review" process, where every year they sum up their current state of affairs, they only briefly mention "risk" in two contexts. One, in a small subset of grants reserved for "high-risk/high-reward" proposals that we'll get to later. Two, in an oblique mention in the description of the role of program officers, who ultimately make the funding decisions, and that if a panel review deems a project as "risky," they can still give the proposer a small seed grant to help get the idea off the ground.[20] The very existence of that comment implies that "riskiness" in science is seen as a detriment, but that the NSF has graciously offered pathways for those kinds of proposals to potentially get something, *despite* the wishes of the panel reviewers. Bless their hearts.

But why are the researchers who make up the panels, and the program officers who allocate the resources, so risk averse? Why don't we live in a world where they *like* risk and want to see more of it?

Because funding agencies like the NSF and NIH have to report to Congress and get approval for budgets every single year, and members of Congress are beholden to their constituents—the folks who vote for them. The funding agencies have to sell their programs as *progress* to the American leadership and taxpayer alike—and progress demands *results*. A bunch of nerds getting all the money they want with nothing to show for it doesn't really *sell well*.

How many times have we heard members of Congress bemoan the "wasteful" spending on "pointless" science projects that seem to go nowhere? Too many times, that's how many.[21]

Also, there isn't enough money to go around.

The numbers sound impressive on the surface. For the past few years, the NIH has commanded a budget between twenty-five and thirty billion dollars for biomedical research (more than twice what it was in the 1990s). The NSF itself spreads about eight billion dollars across various disciplines. With all sources of federal funding taken together, engineering research comes out as a distant second place to biomedical, with typical budgets in the range of ten to fifteen billion dollars. Physical sciences tie for third with "all the biology stuff that *isn't* medicine" around the six-billion-dollar mark. Every other field of scientific endeavor scrapes by with less than five billion dollars a year.[22]

But that's not a lot, comparatively speaking. Federal funding has been trending down (or at best, flatlining, depending on the agency and discipline) for over a decade (in inflation-adjusted dollars, which is what really matters). Despite the bump that the NIH got in the early 2000s, things haven't looked so hot for that agency ever since. Every single other federal funding agency has been on life support for decades, with budgets getting steadily shaved as time goes by.[23] As a percent of discretionary federal spending (and total federal budget, for that matter), research funding has been going down, sitting lower than 2 percent of the total budget since 1982. In a painful irony, as our economy has grown and the country become wealthier, the fundamental science that has fueled a lot of that growth has not share in the bounty: federal research funding as a percentage of GDP has been going down since the 1970s.[24]

Interestingly, research funded by *industry* has been on the upswing for the past decade, stepping up to fill the void that federal agencies are leaving.[25] And they are (literally) profiting from that research. The research that industry engages in is still science—it's still new knowledge about the universe arrived at by a specific methodology—but it's clearly done for a specific purpose: to make products that make money. Yay for them, but unfortunately not the subject of this book. What I'm talking about here is *basic* research—the deep science that explores the world around us for the sake of exploring the world around us.

A deep science that has seen funding cut by 30 percent of what it was at the beginning of the millennium.[26]

Those precious federal dollars—as scarce as they are—are crucial: the vast majority of university research in most fields of science isn't funded by the universities themselves but by national agencies.[27] So if you want to Make It as a scientist, you've gotta play the grant game.

And the grant game is, by and large, conservative. No, not *that* kind of conservative. I mean conservative in the sense that with little to no prospects of future increases in budgets (and a declining share of the wealth the nation produces) and tens of thousands of eager scientists pressing up against the doors and windows like academic zombies, the funding agencies don't want to take risks. It's a conservative *frame of mind*, not a political position.

So to get a grant, you have to be a part of that conservative mindset. You have to be as careful and cautious as possible and demonstrate ahead of time (yes, sometimes before you even get the grant) that you know the results you're going to get. Proposals that get branded with the scarlet R for risk are simply unlikely to succeed in this stressed environment.

Over time, it's getting harder and harder to get a grant. In 2008 the NSF received exactly 33,643 proposals. In 2018, that jumped to 40,364.[28] That's more proposals—and more scientists—fighting over smaller piles of money. Scientists themselves perceive this: in one 2014 survey, a full 83 percent of scientists said that it was harder to get federal funding then than it was five years previously.[29]

We can see the ossification of science—and the lethargy and risk aversion that come with it—in those same NSF Merit Review reports. The average number of collaborators per proposal has been steadily increasing over the past decade, but the median award amount is lower than ever, meaning less money per person.[30] And as we saw, larger collaborative networks make for slower science.

Speaking of slow, the amount of time it takes for the NSF to complete a review is on the rise, now taking longer than six months for a proposer to find out if they made it or not. And since the average award amount is decreasing, that means waiting longer and longer for less and less money—the median award is only enough to pay for less than a month of a typical professor's time. That's right, and I'll put it in italics to really drive it home: *less than a month.*[31]

Overall, across all research proposals in 2018, the NSF could only afford to give 22 percent of them awards (leaving 1,835 proposals ranked "Very Good," representing another $1.5 billion of potential science funding, to twist in the wind). Back in Ye Olden Times (i.e., the 1990s), the success rate was solidly in the 30–35 percent range.[32] But that award ratio is heavily biased toward senior, established researchers. Early career scientists had a dismal 19 percent success rate, while more senior scientists saw a 24 percent success rate. Of all the awards made in 2018, less than 25 percent went to young scientists and their bright new ideas—and overall they got less money. Sadly, this is not the historical norm; the "shift to senior" trend has been going on for at least a decade. In 2009, the NSF supported 33,536 senior scientists and roughly an equal number of postdocs and graduate students. In 2018, that ratio shifted, with around 35,000 old-timers getting awards and only around 30,000 youngsters getting some cash.[33]

The NSF isn't alone in this. I won't even touch NASA and the DOE, as that's just a dismal swamp of grant funding. But even though the NIH received a massive booster shot in the early 2000s, their success rate has actually gone *down* to the same 20 percent level as the NSF. Yes, there's more money floating around, but there are also *twice* as many proposals begging for that money than there were in the 1990s.[34]

Compounding this is the fact that fewer and fewer reviewers are reading over the proposals every year, and as we saw, they tend to be made up of . . . senior researchers.

Also, the statistics reveal something sneaky. While the *median* award amount in 2018 was $140,000, the *mean* was $178,341. For those of you not statistically inclined, this means that the awards are biased toward several high-dollar-value grants, balanced by large numbers of smaller grants given to the vast swath of unwashed masses.[35]

What happens when funding amounts change also shows the bias toward established, senior (i.e., safe) researchers. When the NIH got extra funding at the beginning of the millennium, it largely went to established scientists in large groups—the very people that were already well funded.[36] The same thing happened during the 2008 federal stimulus (or anytime funding is unexpectedly increased)—it tends to go to the people who have the time and wherewithal to submit a new grant application for short-term funding on the fly: senior researchers. And when there are unexpected *cuts* to federal funding levels? The first ones on the chopping block are the young scientists trying to make a name for themselves in the cold, hard world.

In fact, the NSF does have some programs to provide small amounts of funding to risky ventures, like the EAGER and RAPID awards. However, most of these go to engineering and computer science. They represent less than 5 percent of the awards (and even less of the dollars) in physical sciences.[37] "Riskiness" is simply not a priority for our federal funding agencies.

To summarize: senior researchers working on established lines are typically the ones reviewing proposals. They're also the ones typically receiving grant funding. They know how to work the system by writing grants that are "guaranteed" to make progress, so that the heads of the funding agencies can go tell Congress that we're . . . well, making progress. Young researchers have a much harder time winning awards, because they're so risky by very nature of their youth, and risky projects themselves (new methods, new techniques, new research directions, new *anything*) are actively discouraged by the funding agencies.

The whole cycle cements mature fields and established researchers in their places, embedded in large collaborations where all the easy problems have been solved, leaving only the hard—but safe—questions to tackle. There isn't room for new ideas—they've been crowded out by a system that favors proven slow results over uncertain rapid advances. It's a deliberate *choice*, made by scientists over the course of decades. The main reason that science is risk averse is that we have intentionally constructed it to be this way through our grant-making funding structures. I don't blame scientists for making that choice—when faced with an irate senator knocking on their door demanding answers, the easiest way to respond (and make the problem go away) is to give the senator some answers.

Demonstrable progress. Slow but certain results. An avoidance of risk. Scientists have crafted a culture of stagnation in order to survive. But this strategy won't work for the long haul.

I'm not knocking senior researchers here. They are great people. Some of them are even good friends of mine. And I know it seems like the previous discussion has just been an extended, reference-backed whine about how senior researchers are being . . . senior, and taking all the resources for themselves. And you might rightly ask: don't they *deserve* that? They put in the hard work over the course of years, fighting the good fight for the advancement of human understanding, and now that they are full of both knowledge *and* expertise, shouldn't they be running the show?

Sure. As long as enough resources also get funneled to young scientists and to scientists with new ideas, which isn't happening.

I mentioned in the last section the so-called dwell time, the time it takes for a funding agency like the NSF to make a decision. That dwell time has been going up over the last decade, so scientists have to wait longer and longer to find out if they're getting grant money. What I didn't mention is the *preparation* time—the amount of effort scientists are putting in to make the proposals in the first place. Proposals just don't magically appear out of thin air; you have to work (really hard) to make them happen and make them competitive.

And if the acceptance rate for proposals is too low, you enter a death spiral. You spend months preparing a proposal, months more waiting for an answer, then . . . nothing. So you wait for the next cycle, spend months tweaking the proposal, months more waiting for an answer, then . . . nothing. So you wait for the next cycle, and on and on and on.

With the typical amount of time that scientists spend writing grants (a couple months) and an acceptance rate below about 20 percent, many end up *never* getting a grant; they go through multiple cycles of rejection before they quit or lose their bid for tenure.[38]

And even if you do finally get a grant, it only pays for (on average) less than a month's worth of productive research time. So you end up spending months (and even years) of effort *just writing grant proposals*, and never doing actual research.

But good news! While low, the acceptance rates for the NSF is above 20 percent . . . if you're a senior researcher. Not so if you're a junior scientist.

The youngest scientists are caught in a trap. They *have* to apply for grants, because grant money means more time for research, means more papers published, means a higher chance of success. But they're not getting grants and have to reapply and reapply. They end up just writing grants, never producing science, then leaving.

All because young scientists are perceived as untested and risky.

That leaves a system where the only people who win are the senior scientists, because the system is biased in their favor, and they're the only ones with the luxury of enough time to keep writing grant proposals—they don't get caught in the death spiral.

That issue aside, it's perfectly possible for the old-timers to be risky. After all, that's the intention of tenure: they have the job security necessary to boldly try new ideas without fear of not being able to pay the mortgage. But as we've seen, as fields of science grow, they tend to become large and cumbersome—and the leaders of those fields become much more risk averse with time.[39]

Instead, senior researchers either focus on already established lines or take up leadership positions in gigantic collaborations. In the Planck mission, for example, most of the work done to produce the results was cranked out by postdocs and graduate students (who were then unceremoniously discarded once the mission—and funding—ended), while senior scientists called the shots.

The funding agencies are becoming ever keener on funding these giant collaborations, massive missions, and "centers" (big collaborations focused on a particular research line), because the results of those projects are much easier to quantify. For example, out of a total $1.3 billion that the NSF allocated for mathematical and physical sciences,

$90 million went to "centers" and $289 million went to massive facilities, and both of those numbers are going up with time.[40]

It's much easier to say that we will measure some property of the universe ten times better than we did in the previous generation than it is to say that we have no idea what we're going to get.

In some ways, this aversion to risk can be a good thing. It's pulling in funding day after day, somewhat reliably. It turns science into a machine that just keeps pumping out numbers and figures and data. It keeps junior scientists employed . . . for a while, at least. It gives funding agencies something reliable and predictable to report to members of Congress and something reliable and predictable to report to the general public.

But is science supposed to be . . . reliable and predictable? Could Newton or Einstein have reached the dizzying heights of intellectual curiosity that they did in the current scientific climate? Isn't science supposed to be . . . I don't know . . . fun? Serendipity and creativity play a crucial role in the discovery of new knowledge, and the opportunities for random chance and leaps of free, unfettered thought are becoming narrower by the day.[41]

Science is slowing down. Adjusted for the number of scientists performing research, we are learning less and less about the world around us with every passing day. The general stagnation that pervades most scientific fields makes us afraid to continue learning. The emphasis on giant collaborations and huge projects makes us slow, *on purpose*, because the point isn't to push new boundaries of knowledge and strike fearlessly into the unknown—it's to make sure the grant money keeps rolling in.

Slow-but-predictable science makes for an easy sell, but it's the wrong product. We've spent decades trying to convince the public and policymakers that science is *useful* instead of *wonderful*—literally, full of wonders. That the only reason to continue funding science is because it produces useful results, not wonderful results. We've been playing a dangerous game, arguing that science is a sure bet, a safe investment. That sets up the incentive structure that keeps modern science cumbersome and lethargic.

This incentive structure is its own version of a death spiral. The more we convince the public (and lawmakers) that science must be predictable, the more it becomes predictable. And in a sense, it works, because we've constructed the scientific apparatus to continue delivering predictable, reliable results. But those results are increasingly becoming less interesting, less useful, and less worthy. It is, technically, in a narrow and limited sense, progress, but not the kind of progress that we necessarily want. The more stagnant that science becomes, the less enthused the public will be to fund it. The less the public is interested in science, the less they'll want to support it. The less the public wants to support science, the more risk averse we become.

And we end up exactly where we are, wondering how we got here: grant funding on the decline, fields of science petrifying, a public losing interest, and eager new researchers marginalized.

The good news is that it's not too late. The *Titanic* of science hasn't struck the iceberg yet. But change has to come from within. Thankfully, the solution to this vexing problem of ossification is rather blunt and straightforward, but it still requires buy-in from the entire scientific community: simply put, we need to encourage risk and celebrate failure.

It starts with the top of the funnel, with hiring and promotion of junior researchers. The current system rewards success, which is defined—as we saw—by writing a lot of papers. The best way to write a lot of papers is to join a big, established collaboration, or to simply get lucky. So we need to redefine what success really means. It shouldn't mean contributing as a cog in a science-machine or cracking some giant mystery of the universe. Instead, success should be *trying*. About coming up with bold new ideas or approaches. About finding the boundaries of human knowledge and not just inching over the edge but plunging unabashedly into the darkness.

Yes, most of those attempts to forge new creative directions in research will be wrong, misguided, poorly timed, and so on. But we have to remember that even experiments that fail to turn up a result, or involve a brilliant idea that turns out to be incorrect, or the thousands of other permutations that can stop a research direction dead in its tracks, can still produce valuable information. Negative results and even *wrong* results still yield good knowledge. If we encouraged young researchers to try new things, to "move fast and break things," and we rewarded it, we would develop a scientific culture that saw innovation, experimentation, and radical ideas as the foundations on which to build countless inquiries into the natural world.

Second, we need to fundamentally alter the way that funding agencies determine the merit for grant awards. Simply increasing the amount of money that these agencies can dole out won't do the trick because when that happens it tends to go to established researchers. There is indeed too little money going to too many hands, but while I generally support increased science funding, I think we need to take a critical look at the explosive growth in the raw number of scientists, a problem I will discuss later. For now, let's revise the criteria that funding agencies use to make decisions. Currently, less than 5 percent of NSF grants go to high-risk/high-reward projects. Why 5 percent? Why not 10 percent? Why not 50 percent? That percentage of risky projects can be literally anything that we choose it to be—scientists are in charge of our own destiny because we're the ones serving on the grant committees and making the recommendations. Riskiness in a scientific project should be seen as a virtue not a vice, and rewarded as such.

What if we flipped a switch tomorrow and decided to allocate a larger proportion of funding to grant proposals marked as "risky"? What would happen? We might lose some of the giant resource-hogging projects that are the mainstay of modern research. But what would we gain? It's impossible to say because it's been a long time since we've ventured into the unknown like that. It's safe to say that we would have a lot of misses, a lot of projects that end up dead in the water, research trees barren of fruit. But we might also surprise ourselves.

And while we're at it, what if we inserted an element of randomness in the grant-making process? Once a collection of proposals advanced through some sort of basic filter, what would happen if the results were simply randomly awarded like a lottery? Would science be worse for it? Or would it be better, because fewer people would be involved in the selection and recommendation process and all the biases and groupthink and inertia that every person adds would simply vanish?

I understand the desire to craft survey documents and report on the current state of the field in the various disciplines in science. And I understand why funding agencies just love that stuff—it makes their job so much easier because they can just benchmark any proposal and give the community's stamp of approval. But soliciting a survey to get a sense of a field's current state of affairs is one thing (and very interesting and useful); it's quite another to use that as a guide for *future* decisions. How can we look to the future if we're constrained by the past decisions of our colleagues? Funding agencies should examine proposals within the context of the proposal itself and its own self-contained merits and demerits, not referencing what the community at large is up to.

But one of the main reasons many fields have implemented survey documents is to handle the complexities of choosing which large, gigantic, expensive collaboration to advance. So to address that, we need to pivot to encourage smaller, nimbler collaborations. Overly large collaborations should be avoided because they aren't agile enough to compete with risky ventures. Do we really need "centers" with dozens of scientists spending more time networking and less time perusing interesting ventures? Do we really need to take an experiment we did before but this time do it bigger? Young scientists with fresh, new ideas need to be celebrated as the game changers they can be, not as human computers serving some massive lab or mission.

And now for something radical: maybe we should end the tenure system. It's not doing the thing that it's supposed to. It's *cementing* lethargy, not discouraging it. What's the point of fighting for tenure and a secure position when you're not going to do anything interesting with it? Maybe academic science should be a little bit scarier from day to day. You know, like a normal job. Maybe senior scientists would be more willing to take risks, try new research lines, and come up with genuinely interesting ideas if they had to constantly, year after year, fight against junior scientists entering the field with thoughts of their own.

All of this won't work unless we get the public on board with what science actually means. Science isn't merely a collection of interesting facts about the world (although those facts can be pretty awesome); it's a *method*. It's an approach to gathering knowledge, not the knowledge itself. And that approach can take many forms: slow, incremental, cautious steps forward, or far-reaching, risky leaps.

We need to educate the public, through vigorous public outreach, about the fundamental aims and methods of science and the fundamental joy of discovery and trying new ideas. Just like the required shift in perspective for hiring and promoting junior scientists, we need to celebrate and promote the *attempts* rather than the *results.* We need to convince the public that science is at its most interesting when we don't know what we're going to get. When it is about exploration and discovery, not refinement of known results and ensuring grant success. When it's about the unknown. When it's about the weird. When it's about the wild. When it's about the . . . wonderful.

Since the funding agencies must answer to policymakers, and the policymakers are ultimately beholden to the public, a changed public perception will liberate the grant-making progress, allowing it to become what it's supposed to be: a propellent, a catalyst for accelerating discovery.

Funding agencies have to work hard to make a case that fundamental research is just that—fundamental—and may not lead to useful results . . . or anywhere at all. Hopefully, with a public supporting the institution of science rather than the product of science, scientists themselves can stop asking, "What project is likely to be funded?" and start asking, "What project is most interesting?"

There are more questions in the universe than we can ever possibly hope to answer. Science can be daring, audacious, and bold, and it's in those directions that we have the most to gain—the highest potential for a revolution in our thinking. But to get there, we first have to accept the reality that failure *is* an option.

3

The Lack of Careers

I GET ASKED A LOT OF QUESTIONS. What's at the center of a black hole? What came before the universe? Why don't we understand quantum gravity? You know, the usual gamut of space-related queries. But there's one question I absolutely dread getting. The one question where I can't give an (honest) answer. The one question where I don't have some pithy summary at the ready. The ones question that stumps me, every time:

How can I be a scientist?

One of the reasons I get this question all the time—and I mean, *all the time*—is that we, as a society, keep telling our kids that science is *cool*. But not in a "your friends and peers will admire and emulate you" kind of cool, but in "all the adults agreed that we need more kids taking up careers in science" kind of cool. The latter kind of cool is decidedly less, well, cool than the first, but it's still the message that kids get. And kids being kids, a thousand times more perceptive and intelligent than we ever give them credit for, they pick up on the underlying message: if you have any interest at all in any corner of the human experience that intersects with the skill set necessary to become a scientist, then that's what you should do, because it will make all the grown-ups happy.

Naturally, some kids are just going to be interested in science from the get-go. Some will come to it later, as a part of the twisting and winding path through life that everybody takes. But we're also incentivizing kids (and especially incentivizing their parents) to encourage careers based on science and engineering. We've convinced ourselves that we need more scientists and engineers running around in order to make the world a better place. Who will solve all the problems caused by the current generation of adults? Why, certainly not them—they've got better things to do. So surely the *next* generation can dig us out of this mess that we've made. Right?

I mean, part of this is true. We do need a next generation of scientists and engineers, because once the current ones retire, how will we learn new things and build new stuff? But there does seem to be an unnatural, society-level *push* to encourage kids to come into the folds of science and engineering. To feed their nerdy, curious little hearts.

Despite countless attempts to have a prepared answer at the ready, I still stumble over my words and muddle through a response every single time. Part of the reason is that it's so difficult to clearly articulate what makes a successful scientist *successful* (besides the horrible metrics that we've come up with). The scientific career is one of the strangest paths one can take in modern society. The career experience for most scientists looks more like a cross section of medieval knights and celibate monks than it does, say, a financial analyst or elementary school teacher. I have a hard time describing it in a quick-response Q&A format because *it's literally hard to explain.*

But mainly I have a hard time answering this vital question because my heart breaks when I hear it. When some bright, precocious, adorable kid or some eager, smart teenager or some passionate, earnest adult asks me this question, I have two choices. I can tell them about math classes and undergraduate experiences and independent research and all the other facets of scientific training, and if they stick to their guns and work hard they can carve out a career for themselves as their very own explorer of the universe around us.

Or I could tell them the truth.

I could tell them that there are no jobs. I could tell them that they will spend tens of thousands of dollars and sixty-plus hours a week, sacrificing stable family lives and financial security, through the course of a decade, all to leave the field and their passion behind for a day job they weren't trained for.

I could tell them that for every open tenure-track faculty position focused on research in science opening around the world, there are dozens of PhD candidates produced.

I could tell them that their colleagues and mentors will set up a toxic competitive environment, where if you're not willing to work yourself to the bone, there's another dupe around the corner who is.

I could tell them that their superiors will not train them for work outside academia, let alone *talk* about it, knowing full well that they likely will not move past their current position.

I could tell them that the whole thing is a racket, a pyramid scheme, structured to benefit the full-time faculty currently in control of the scientific world, with no regard given to the students and junior colleagues who actually do the bulk of modern-day research. That the most they'll offer is empty platitudes, well-wishes, and attaboys (even to the girls, but that's another chapter).

I could tell them that we don't really need more scientists—that we have too many already.

But I can't tell them the truth. I don't want to be the one to break a little kid's heart, or to be the wet blanket that puts out the fire of someone's dreams.

And so I'm telling it to you, and maybe we can change the system so we don't have to lie anymore.

We have a journey to go on. These are the trials you must face in order to become a scientist in the twenty-first century. If you want the prize, which usually means a tenured position at a major American research university, then this is exactly what you have to do. There are of course many other paths to science, working for nonprofits or corporations, or pursuing research at institutions abroad. But when it comes to fundamental research that uncovers the inner workings of the universe, with the search itself the primary goal, you have to aim for the American university: they perform the vast majority of fundamental science worldwide. So this is the path you must take.

Welcome to the Gauntlet.

The first stage of the Gauntlet happens in your chosen college. We put a lot of pressure on high school kids to pick their careers as early as they can, which is a mistake. Do we really expect fifteen-year-olds to understand the shape and structure of the entire rest of their professional lives? Let's assume you did decently well in high school, have a good math background, and you've entered college with a dream of becoming some sort of scientist.

It's no secret that the university experience in the United States is too expensive; tuition has increased over 1,000 percent since 1963, well over 2.5 times the inflation rate.[1] A college education is now the second most expensive thing that most people can buy, after a house, and most of it is paid for through student debt. We're asking teenagers to take on that massive amount of debt before they're even considered responsible enough to drink alcohol. But that is just the tip of the iceberg.

After four or five years of classes, independent projects, and late-night study groups, you're going to have a little piece of paper that claims you're educated, and a bill that is likely much higher than your entry-level salary. Entering the workforce (or graduate school) with that kind of debt severely limits options and maneuverability, which is a general ill that is afflicting our modern society, and it applies to budding young scientists as well.

And I hope you picked a good school. But "good" means a lot of different things to a lot of different people, and what society has deemed to be "good" may ironically not give you the best education. For example, according to various polls or college rankings or your local guidance counselor, you might think you should go to a top-tier school that's high on the various lists of rankings. You might even think that you will get a quality education from the scholars at those leading lights.

But you are much, much more likely to be taught by a graduate teaching assistant whose only purpose in being there is to begrudgingly pay for their own graduate education. Or you will be taught by an adjunct professor, a part-time underpaid teacher.

Most importantly, almost none of the people teaching you are trained in . . . teaching you. Even if you get into a class taught by Prof. Famous, their training and expertise are rooted in research, not education, and the only reason they're teaching you is because they couldn't use grant money that year to buy themselves out of the obligation. All this *despite* the increasing cost of education—you're not getting more educational bang for your borrowed buck.[2]

If you're interested in quality of education for your science career, your best bet, ironically, is an institution that does not offer PhDs. No graduate teaching assistants to do the work that the professors are supposed to do—the faculty are the ones actually in the classroom. On the other hand, those top-tier research universities do offer something: access, which you will find as you go forward in your career is your most powerful currency. Students from less well-known schools have a harder time "breaking in"—the largest, most powerful universities have an incestuous habit of simply circulating students and graduates among themselves.[3]

So you have to pick: a better education or better connections. Choose wisely.

But anyway, as you continue to pile on the zeroes to your student loan debt (and the cost of education keeps rising with every passing year),[4] at some point you get introduced to the mechanics of your chosen field of science. For the first couple years, it's all introductory classes—the basic ins and outs and fun and math of any field.

But as you inch closer to graduation, you start to get a taste for the competitive atmosphere. Sure, there's always in-class rivalry as the top students compete for the best grades, but beginning in your junior year you realize why they care so much: in order to get into grad school, you need letters of recommendation. And the easiest letter of recommendation is from a teacher in a class that you excelled in.

"The applicant performed better than over 95 percent of all the students of Introduction to Field-Specific Topic that I have ever taught in my career" begins almost every single grad school letter of recommendation.

Making this competition more difficult is the fact that, despite the cost of education continuing to skyrocket, there are more students enrolled in undergraduate universities than ever before—and that too keeps going up. I don't dare get into the economics of all of this, and for our purposes we don't really need to know *why* this is happening, but we need to grapple with the fact that it *is* happening.

Undergraduate enrollment has exploded in the past twenty years in every field from engineering to agricultural science.[5] Just as an example, in 2000 there were 18,597 freshly minted bachelor's degrees in the physical sciences. In 2017 there were 30,730. Across all of science and engineering, the number of degrees awarded every year has followed similar trends, nearly doubling to over 700,000. This is *great* news if you're a university department. More undergrads means more tuition dollars flowing in, which means more cash for whatever you want to do—the amount of tuition income has tripled in twenty years. It's not often that you get to be in a position where your biggest challenge is working out the logistics of how to unload all those dump trucks full of cash. But as we will see, that money is not necessarily allocated to more or more effective teaching, or more long-term research positions.[6]

But as a student, this is bad news. You get a lower-quality education, reduced access to those famous researchers, and a *lot* more competition. All those friends you make in the first couple years become your rivals, chasing after those precious letters of recommendation and those even more precious research appointments.

Ah yes, undergraduate research. The thinking goes that if you want to grow up to be an independent scientific researcher, then you should do research as early and as often as possible. It sounds innocuous enough, and maybe even a little bit smart (and there are certainly a lot of university-based publications touting the benefits of it), but it only makes sense if the number of undergraduate research opportunities keep pace with the growth of undergraduate students competing for those opportunities. They don't. And so, as usual, what once started as a reasonable way to discover and prepare promising young students becomes yet another unreasonable competition, this time inflicted on a population just barely considered legal adults—any younger and this practice would probably be outlawed.

It's *possible* to get into grad school without doing undergraduate research, the same way that it's *possible* for all the air molecules in your room to suddenly collect into a corner, leaving you in a vacuum. You just might have to wait through many multiples of the current age of the universe for it to happen.

This is toxic competitiveness: an environment where blind luck and established connections, not the things we're actually trying to capture like research acumen and personal drive, lead to success. We insist that undergraduates get some research experience but then don't give those undergraduates research opportunities, and we sit around wondering why things are so broken.

The last thing you'll have to face in this first stage of the Gauntlet is a test. No, I'm not speaking metaphorically—I mean a real, actual test. Some sort of Big Annoying Test, like the SAT but harder. For physics, one has to take the physics portion of the Graduate Record Exam. And yes, your score matters. Some universities state clearly on their website that if you score below some arbitrary threshold, then you shouldn't even bother applying. Some are a little vaguer. But all use the score to filter out candidates, pretending that how you perform on a multiple-choice test in any way reflects your abilities as a scientist.

I need to note something here. Universities often say that they don't pay much attention to GPA or some test score, or undergraduate research experience, or references, or the repute of your institution. But according to an analysis of nearly 400,000 self-reports from graduate applicants, they do.[7]

You might not be surprised to learn that the entire research world grinds to halt at the end of every year as everyone and their advisor is busy (a) writing applications, (b) writing letters of recommendation, (c) reading applications, or (d) reading letters of recommendation. Research productivity plummets every year as everyone breaks out their CVs to do the dance, hoping to get into some better position, or being on the other side of the table and dealing with the usual flood of applicants.

You'll get your first taste of that annual harvest cycle in undergrad. Let's say that you've done well in your classes. You've had some internships and undergraduate research experience. You've got a great personal story. You did well on the big exams. You have drive, ambition, and skills. You have letters of recommendation. You've got the whole package. You put your life on hold and submit to a dozen grad schools.

What happens now? Certainly some students never intended to have a long-term career in science and are totally happy to take their degree out into the world, making friends and making money. But every year hundreds of thousands of students apply for graduate programs across the country, all fighting over a few tens of thousands of open positions.[8] Graduate admissions committees have all sorts of rubrics and metrics that they pretend are useful, but largely it's about luck and positioning: assuming you pass through the grade- and test-based filters that admission committees lie about not paying attention to, the rest of the process involves having the right connections, the right names to drop, and the right institution on your diploma.[9]

Despite all this, interest in academic careers remains high,[10] which means that once you get into grad school the pressures don't ease up. They only get worse.

If you thought the undergrad portion of the Gauntlet was tough (and it is), it's *peanuts* compared to what you'll face in graduate school.

Now, some of the intensity of the grad school experience is explicitly placed there to serve several useful purposes. We want not only bright people working in science but also *motivated* people. It's simply not good enough to be "smart" (in whatever definition of that word you choose)—you have to be smart, persistent, insistent, and determined to succeed in the constant war for knowledge against nature. And so grad school is tough on purpose, to weed out the unwilling.

That's . . . fine, to a point. We want a PhD to actually carry some weight and not hand them out to anyone willing to write a check for a hundred grand. I think I can speak for all of human society when I say that it's A-OK to make advanced scientific training just a little bit demanding.

But there's a big difference between "let's make this part of the experience rough because we want to polish some rocks into gemstones" and "this part of the experience is accidentally insane because we set up some perverse incentives and have no idea what we're doing."

Your first problem you'll have to face as a science graduate student is money. Education costs money (a lot of it), and it's not easy to find. You just graduated from college with a bill attached to the bottom of your diploma, and those numbers are only going to get higher.

So you have a few choices. You can pay for graduate school, hoping to reap the financial rewards in the future (good luck with that). About 40 percent of graduate students across science and engineering end up having to take this option—taking out loans or working their tails off, spending even more tens of thousands of dollars for that degree that will never, ever pay off unless . . . they leave science and engineering and make more money outside research.[11]

You can get a fellowship, having some private foundation or national funding agency foot the bill. But they're about as competitive as competitive can be, so don't count on it.

Many science graduate students enter as teaching assistants, or TAs. With absolutely no training, no expertise, and no practice, they are handed a guidebook, dumped in front of a roomful of eager young undergrads, and told to "teach some science or whatever." In the land of the blind, the one-eyed man is king, and so I guess technically these TAs are more well versed in their subject matter than the students sitting in front of them, but it's not exactly by a lot.

Or you can find an advisor who will hire you as a research assistant (RA), allowing you to learn the ropes of science in a one-on-one, apprentice-style capacity.

Let me say this another way: you *must* find an advisor to hire you as an RA, otherwise your career is stillborn. If nobody wants to take you on, you'll be quietly and discreetly shown the door. If you want to work for a professor who doesn't have any funds to hire you, then either (a) that professor is probably not the most productive member of their field or (b) that professor doesn't have time for you and you'll have to teach part-time for your entire graduate career, limiting your ability to engage in useful scientific research—and limiting your prospects post-PhD.

I saw examples of both of these at my grad school. I saw some of my fellow students spend half a decade as a TA for the opportunity to "work for" Nobel Prize winners, only to have hardly any interactions with them and a lame form letter for a recommendation. And I also saw students work with an underfunded professor and not have any opportunity to advance.

I'm talking a lot about advisors because the choice of mentor is, by far, the absolute most critical decision you will make in your scientific career.[12] Your advisor will (hopefully) take you under their wing, teaching you all the little details of actually *being* a scientist that you simply can't communicate in a classroom setting. They'll teach you how to write papers, how to formulate problems in your field, how to interact with collaborators, how to actually construct your day-to-day routine in existing as a modern scientist.

Within a year or two at most, you'll phase out of taking classes (yes, there are classes and massive exams too, adding to the pile of simultaneously starting a career in research and teaching your subject to undergrads—that's part of the "boot camp but for nerds" construction) and focus on research as a junior colleague to your advisor.

This can be a beautiful system and one of the things that makes science so unique in the modern world. It really is an apprentice-based organization, where you learn more practical experience and skills from your mentor than any college class or on-the-job training could provide. It's these one-on-one relationships that transform you from starry-eyed wannabe into jaded expert, from student to collaborator, from candidate to scientist.[13]

But it's in this very same system that you learn the modern scientific training regimen isn't really about properly training the next generation; it's about providing grist for the mill.

A typical professor can have anywhere from one to ten graduate students working for them at any one time, and they will graduate a dozen or more students over the course of their careers. As we will see, essentially zero of those graduate students will go on to get tenured faculty positions of their own. The professor gets research done, papers written, grants won, accolades received, awards granted, the whole pile of scientific rewards—in their name. The graduate students . . . do work. And then they leave.

This might be okay if professors were totally up front about the structure and you knew what you were getting into: an opportunity to wave around a PhD in other people's faces for the rest of your life (which is pretty satisfying) in exchange for half a decade of intense work. But instead, faculty actively and explicitly recruit students with academic career ambitions, believing that those are the kinds of students that make the "best" research assistants.[14] And maybe they're right—but those are exactly the kinds of students who want to be professors of their own . . . and don't get to be.

Your mentor is also the person who completely and totally dominates that same early career and completely and totally dominates the outside community's perspective of you. You *have* to keep your advisor happy if you want your career to take off. It's their word that the rest of the community relies on. And since one professor will typically mentor multiple students, knowing full well that not all of them will be able to continue in a scientific career, they're perfectly capable and willing to pick winners and losers among their cohort, flagging some of their students for success and others for . . . something else.

As with any professional structure that features such intense power imbalances, it can lead to bullying, harassment, and worse—but that's the sad story of another chapter.

You'll also first discover as a graduate student the bureaucratization of science, as we explored earlier. Modern science isn't really about discovery and adventure; it's a *business* whose primary job is to sell new knowledge to universities and government funding agencies.[15] You will find, as you pick among possible advisors, that your scientific creativity is severely limited. You won't get to explore in whatever direction most piques your curiosity. You won't get to follow your decade-long dream to study whatever it is you want. You might, if you're lucky and at a large enough university, get to pick from a limited menu of possible research projects, suited for grad student grunt work (in other words, the kinds of timid projects that funding agencies just love).[16]

But pick you will, likely toiling at some obscure corner of some gigantic, collaboration-driven project. You will spend the remaining years of grad school doing science: conducting research, writing papers, giving talks, falling asleep face-first into your coffee mug, the whole deal. And you'll face a lot of challenges along the way.

What does it take to finally get a PhD? A scientists' favorite word to describe this phase of the Gauntlet (and my favorite word too when I'm feeling nostalgic) is "grit." Determination. Perseverance. We need scientists who are relentless, unforgiving, and dedicated.

When I'm not feeling nostalgic and instead looking at raw numbers, what we reveal is a process where 76 percent of survey respondents report working more than forty hours a week. A system where one-third of grad students have sought help for anxiety and depression due to the school experience. A structure where 40 percent of students report dissatisfaction with their work/life balance (in case you didn't get the point by now, there's no room for life in the work of grad school).[17] A structure that suffers from an attrition rate of up to 40 percent from burnout, frustration, and abusive mentors.[18]

And you'll also find a perversion where, despite all this, over half of grad students still want to pursue a career in academia *and believe that they will succeed.*

So after all this work, all this toil, what do you get in return, besides the actual diploma?

Well, you're likely to have an insane (and potentially never recoverable) debt, especially if you're a psychology or social science major.[19]

As you begin to explore post-postgraduate options, you'll find that a career in industry pays significantly more than academic appointments but that nobody in your department is really talking about that.[20]

And when anyone senior in your department *does* talk about life outside academia, it's often described with contempt, derision, scorn, and dismissal.[21] Your superiors aren't just ignorant of life outside the quad; they disdain it—and they'll discourage you from going in that direction.

You'll have, on average, six to eight years shaved off your life, devoted to obtaining that PhD.[22] And keep in mind, that's *in addition* to the four-plus years it took just to get your bachelor's. That's more than a decade of your life

If you do decide that the academic life just isn't for you, then that means you're more than half a decade behind any of your peers in the world of Real Life in terms of career advancement, retirement savings, and likely personal relationships (not to mention a family).

Viewed in a chartable light, the grad school phase of the Gauntlet teaches you valuable skills beyond the actual scientific training. You learn rational analysis. You get proficient at mathematics (whether you like it or not). You get proficient at computer programming (whether you like it or not). You become a certified Independent Researcher, capable of charting your own course through the landscape of questions available. Of seeing difficult problems, posing targeted questions, and designing a program to answer those questions. Your mental blade gets sharpened to unimaginable precision.

You become a scientist.

You might think that now, finally, exhaustedly, you're ready for a career in scientific research. For a tenure-track faculty position or a choice position at a research lab.

You would be wrong.

In 2019 the colleges and universities of the United States printed more than forty-two thousand PhD diplomas for scientists and engineers.[23] Compare that to the less than six thousand awarded in 1957.

Good news! We're producing more scientists than ever before.

Bad news! We're producing more scientists than ever before.

It's hard to tell exactly where all these newfound Dr.'s wind up, which is itself a little discouraging. Universities aren't exactly . . . forthcoming . . . with data they collect about the long-term outlooks for their graduates, assuming they even bother collecting any data in the first place.[24] Doesn't that strike you as just a little bit odd? Wouldn't universities *want* to track this, so they can put all sorts of large numbers and success stories in their glossy brochures, marketing the Rich and Fabulous Lifestyles that await you if you would only come to them for your graduate education?

But they don't track it. They don't do surveys. They don't brag about how successful and well-adjusted their numerous science graduates are.

The reason that colleges and universities don't like talking about the long-term outlooks for science graduates is that it's not fun to talk about, and probably if they did talk about it more, fewer students would sign up.

The best we can do is rely on nationwide surveys, which are themselves a little murky. For undergraduate degrees, the options are relatively simple: you either continue into graduate school or you don't. But the options for PhDs are much richer, with many branching options: teaching, academic research, industry research, short-term positions, long-term positions, regulars jobs, and so on.

Here's what we can see:

In 2019 around 30 percent of finished doctoral graduates reported having no definite employment commitments of any kind, of the temporary or permanent variety.[25] That's right: three out of ten science graduates in our country, some of the smartest and brightest human beings on the planet, with a decade of training in the field of their expertise, *don't have a job lined up when they're done.*

Another healthy percentage (between 20 and 25 percent) go into industry, where they generally find reasonable working hours, high pay, and healthier work/life balance (and largely not doing any science)[26] or head for a government or nonprofit lab.[27] The remainder find some sort of academic appointments, half of which are for teaching positions.

You might have been tempted, before you began reading this chapter, to think that those who remained in academic research, around 25 percent of all recent grads in science and engineering, would find themselves, after suffering this decade-plus intellectual boot camp, placed in a tenured (or at the very least tenure-track) position at a college or university, ready to start a life of doing science and teaching kids and sitting on committees.

Oh, my sweet summer child. Those who've made it this far in science have only reached the third phase of the Gauntlet: the postdoc.

"Postdoc" is one of those weird words that only ever appears in academia and has to be constantly explained to the rest of humanity. It's short for "postdoctoral" and generally means a postgraduate research position that lasts a limited amount of time, usually one to three years. In other words, it's a science temp job for the highly educated.

Let me briefly flip the script a little bit here: surveys show that less than 20 percent of recent PhD earners move to a permanent position within three years of graduation.[28] At any one time, about 10 percent of all human beings who earned a PhD from an American university are employed in these temporary postdoc positions. And many of these postdoc positions aren't even paid gigs—like, some of the smartest and most well-trained people on the planet are just *interns*, and the scientific system is *totally cool with that*.[29]

As with all things in the modern science world, the postdoc started off with good intentions. As a grad student, even a senior one, you're still under the wing of your advisor, leaning on their mentorship and wise counsel (if you're lucky) to grow yourself to independence. But the larger community of scientists isn't exactly sure you're ready for unfettered, independent research. So we construct a sort of staging area, where you're on deck for a faculty position but really seeing for yourself and us if you're good enough to go the distance.

Plus, being a faculty member is tough. You have a lot of responsibilities that have nothing to do with research, and so we construct these temporary positions so you can maximize your research output without all the distractions like committee meetings and grading exams.

Double-plus, it lets science mix around the intellectual gene pool a bit. I haven't mentioned the mobility factor yet, but if you want a career in science you have to be prepared to move around. A lot. One institution for your undergrad, another for grad school, another for a postdoc, maybe even another for a second postdoc, and then the place you can finally call home as a professor. If that doesn't sound like so much fun—not having any stability in your life, let alone minor things like a house and a family, until your midthirties *at the earliest*, well, maybe the Gauntlet isn't for you (and you're likely to not succeed in academia).[30] But anyway, ideally, your postdoc research group and advisor will be *way* different than your grad school mentor, giving you different experiences and skills.

That doesn't sound so bad, right? Back in Ye Olden Times (i.e., the mid-twentieth century), this wasn't a bad setup. You got out of grad school, did a two-year stint with another research group, then slid into a faculty position somewhere. You got bonus training and expertise, a couple years to only focus on research, and the chance to live somewhere fun without having to commit.

But over the past several decades, the postgraduate options for young scientists have steadily shifted from permanent faculty-level positions to more and more of these kinds of temporary appointments.[31] The postdoc has transformed from a painful-but-legitimate part of overall scientific career training into hell's own waiting room, a way to exploit as much scientific labor out of bright young minds as possible without having to go through the trouble of actually promising them anything permanent.

Yeah, sure, there are *technically* more science jobs in the United States than ever before, but they also *technically* lead to absolutely nowhere. The postdoc is no longer a streamlined pipe that runs from PhD to tenure but just another way for the scientists at the very top of academia to keep the grant dollars rolling in without giving their junior colleagues any pathways to a career of their own.

All told, almost all tenure-track positions across science expect at least one, and sometimes two to three, postdoc appointments before a candidate is considered "worthy" enough for a permanent position. I mean this in the most brutally efficient sense: in most fields of science, if you don't have two postdocs under your belt, faculty hiring committees *won't even look at your application.*

All this work, all these years and tears poured into the Gauntlet—undergrad, grad school, multiple postdocs—all for the *chance* to *maybe* enter into a permanent research position.

And it's still not over, and even the Grand Prize, a tenure-track position at a major research university, isn't exactly what it seems to be.

Despite the growth of postdoc "training" (we'll use that word loosely and charitably here) positions, at the end of the day the number of PhDs produced far outstrips the number of open faculty positions. While the total number of full-time faculty positions has tripled in forty years from 100,000 to roughly 300,000, the number of PhDs produced annually in that same time frame has *octupled*.[32] In our current situation, we're seeing a funnel going from hundreds of thousands of undergraduate science degrees, to tens of thousands of graduate science degrees, to thousands of open faculty positions. It's difficult to track the full postdoc-to-tenure pipeline because this can often take the better part of a decade, but estimates range from 10 to 20 percent of all PhDs eventually finding themselves in a tenure-track post at a university (where they will likely spend the majority of their time teaching, not doing research), only a small single-digit percentage making academic scientific research a full career.[33]

So we have a system here that churns out doctorates, keeps them in the spin cycle for half a decade or more as they gain "valuable science experience" as a postdoc, and then *spits them out anyway.* A system where they hop around the country (or world), sacrificing every sense of stability and normalcy as they fight through the Gauntlet, only for them to discover that there isn't even a pat on the back as they're walked out the door.

This is what I'm talking about when I keep bringing up the toxicity of the science career advancement system. It's hypercompetitive, with a winner-takes-all incentive, and it *still* doesn't even make any real winners.

In 1973, 90 percent of science and engineering faculty were employed full-time. By 2013, that number dropped to 70 percent.[34] So even for the beleaguered scientists who do manage to achieve a faculty position, nearly one-third of them still have to pick up work as a barista or day laborer to make ends meet. Indeed, 73 percent of *new* faculty positions are no longer tenure-track, a decline from 50 percent just in 2015.[35]

It didn't start out this way. Originally, adjunct faculty were an option of last resort, when you needed someone qualified to teach a class but were severely short-staffed and couldn't count on your normal ranks of professors to do the job. So you looked around for nearby PhDs, stuffed a couple thousand bucks in their pocket, and pointed them at the nearest classroom. A quick, easy solution for nagging little scheduling problems.

Now adjuncts are the norm. They're cheaper than full-time faculty. You don't have to pay them full-time. You don't have to give them a retirement account or health benefits. You don't have to deal with them wasting their time with nonproductive activities. You can hire them on a semester-by-semester basis. They are fully qualified individuals who are willing, and able, to work for relative pennies. As universities continue to burst at the seams with undergraduate students, *someone* needs to teach them all. But since nobody wants to make the long-term investments needed to create more full-time positions, and there are *way* more graduate students coming out than can possibly fill those positions, then the adjunct position is the perfect short-term Band-Aid to slap on the open, festering wound of science career trajectories.

Plus, full-time faculty have the option to use grant money to "buy" themselves out of teaching. They pay the university a portion of their federal funds, and the university turns around and uses it to hire an adjunct in their place (and helps itself to a little off the top). Everybody wins: the full-time professor doesn't have to let teaching get in the way of research, the university gets some extra moolah, and the adjunct professors . . . uh . . . get a job, I guess?

Hooray, you get job! Too bad it's only for this semester, with an hourly rate that would make your local teenaged burger flipper turn away in disgust and a workload that prevents you from taking on any other jobs. And if you're lucky and play nice, *you might get to do it again next semester.*

Check this out. In 1973, full-time faculty aged sixty-five to seventy-five amounted to only 2 percent of the total professorial class. Today it's 12 percent, and 25 percent in research-intensive universities.[36] Think about that: there are so few full-time positions that at high-profile research universities a quarter of all their tenured professors are eligible to receive Social Security checks.

If you got your PhD before 1995 and are currently in a faculty position, chances are you're tenured: 70 percent of pre-1995 PhD faculty are in those kinds of precious positions. For post-1995? That drops to less than 35 percent.[37] In the biological sciences alone, in 1973 over 50 percent of faculty appointments were tenure track. That's now down to a meager 15 percent.[38]

Besides the adjuncts, there's another kind of temporary research position. This one goes by various names and titles, so for our purposes we'll just call it "research associate."

A research associate is just like an adjunct . . . but you do research instead of teaching. You get hired by some group or full-time professor, paid for out of their grant money, and do work for them. It's pretty much just a postdoc without a definitive end date.

The bright side of this is that 40 percent of scientists in any kind of academic position whatsoever report that "yeah, sure, I do research," which is up from 20 percent just a couple decades ago.[39] The darker side of all this is that there is a stunning rise in no-first-author scientists, as in researchers who never, ever claim a research paper as their own, because they're stuck doing the inglorious grunt work in support of some powerful group or giant collaboration.[40]

Those same collaborations that we saw as a product of the timidity and risk aversion of scientists (and the people paying for scientists) also give rise to an underclass of forever-junior colleagues. Researchers who spend year after year never quite knowing if it will be their last, but certain that they will never rise up the food chain and secure a permanent position for themselves.

All that's left, after the grueling Gauntlet of undergrad, grad school, postdocs, and possible temporary research or adjunct appointments, are the winners. The people who made it. The tenured and tenure-track faculty positions. The Scientists.

In case you've missed it already, "tenured" means you can't get fired. It's a job for life, assuming you show up to teach your classes (if you can't buy your way out of that) and commit no egregious moral or legal sins (if you can't buy your way out of those). It's the ultimate prize, the holy grail, the gold medal of a career in science. If you're a tenured professor at a major research university, you've *made it*.

Except, not.

"Full-time" *sounds* awesome—an actual, real job doing actual, real science, but that's not even the half of it. Or the tenth of it. Those two little hyphenated words hide a lot of uncertainty and even more lack of career advanced. "Full-time" doesn't always mean tenured, or it might mean tenured at a university primarily geared toward teaching. If this is your dream in life, then you're living it—but preparing to teach the next generation scientists is *not* what the scientific career training is about, so you end up here because either (a) you've always wanted to do this or (b) you didn't have any other options. You certainly did not end up in a full-time teaching role because the academic career system guided you here.

In fact, the vast majority of full-time faculty—74 percent—do *not* report research as their primary activity.[41]

At the top of the pile is an extremely elite group that, for lack of a better phrase, hogs it all. They get the vast majority of research funds. They are capable of hiring the most underlings. They are able to publish the most—not because they're smarter or more qualified, but because they're in a more luxurious position to generate more papers.[42]

But even among them, things aren't so great. Many universities will explicitly hire more faculty than they need, sending them on the tenure track but only intending to make one of them permanent.

Right, that's the difference between tenure and tenure-track. Tenure-track means you *might* be a full professor someday. You'll get some finite amount of time and a set of goals. Goal #1 is to bring in *x* amount of dollars in research grants. Goal #2 is to write *y* number of papers. Goal #3 is to mentor *z* number of students. Goal #4 is to serve on various committees in your department. Goal #5 is to not royally screw up the teaching part.

When the time runs out, your department sees if you're up to snuff. If you've accomplished all these goals, you're in. If not, you're out. It's just that simple. It doesn't matter if, at that point, you've devoted two decades of your life to a scientific career, earned way below what you can outside of academia, and lived in five different cities.

You're. Out.

Even *then*, sometimes tenure doesn't even come with a salary anymore! The university won't actually pay you—they'll be kind enough to give you an office and an internet connection, but really your fancy faculty position is just a license to beg for grant money from the federal government.

And even *then*, the majority of your time as a tenured faculty member of a major university is not going to be doing science. A professor is more of a middle manager, someone who directs and organizes research performed by grad students and postdocs, not someone who's digging into the code or the lab or the observations themselves. A typical professor spends a significant portion of their time writing grants, teaching, and serving on committees (and presumably maybe also now finally hitting the dating scene and trying to start a family).

I'm not trying to imply that the People at the Top, the full-time faculty who decide the fates of their underlings, are some sort of conniving gluttonous evil monsters. Almost all of the researchers that I've met in my career have been decent folk. All of my advisors and mentors, from undergrad on up, have been kind, caring, intelligent, considerate, earnest, hardworking people. Some of the best human beings I've ever had the pleasure of interacting with. They all cared about me, wanted me to succeed, and truly enjoyed teaching and guiding me and all their other students.

But at best—and too bad if I'm hurting anyone's feelings here—the full-time tenured faculty at major universities are complacent about the problems in science, if not complicit in creating them.

Those senior researchers grew up in this system and survived. They ran the Gauntlet and came out on top. So of course they assume that in order to succeed young researchers just need to do what they did—ignoring the impact of all sorts of factors outside of their control, most importantly raw luck and connections, that led to their current positions. The academic life is the only life that they know, and at first glance the experiences of their junior colleagues don't look much different than what they had to go through. Sure, there may be an extra postdoc position or two involved in the process, but that's no big deal, right?

They do know that they hire more graduate students and postdocs than could possibly fill the tenured ranks of all the universities in the world. But they're also trying to survive themselves, by getting grants and writing papers and getting grants again—and a small army of postdocs and grad students is essential to make that happen. They simply don't have the time themselves to do science; they're too busy with all the other aspects of professional academic life. In other words, I'm not personally blaming every single senior faculty for doing what they're doing—hiring too many students, not helping them explore other options. They're stuck in the same system as everybody else.

Why do we have this system? Who asked for this? Why should a typical young scientist have to endure a half decade of graduate school plus postdoc appointments—we're talking up to and over a decade of this meat grinder—before being considered "worthy" enough for a permanent position?

Follow the money. The United States is experiencing an explosive growth in all undergraduate majors, fueled by cheap loans and a promise of a better future. Those swelling ranks of undergrads produce more people interested in graduate school, more money available for graduate appointments to teach those undergrads, and more people willing to consider a career line in scientific research. Universities are absolutely not interested in turning off that pipeline.

At the other end of the spectrum, the funding structure in the United States favors competitive short-term grants, like we saw earlier when it came to risk aversion.[43] Nobody is willing to invest in scientists or their curious little projects on a long-term basis, and so the shorter the better. This is *great* for existing tenured faculty. They get paid their salaries from their institution no matter what, and get to use that short-term grant money to pay for as many little minions as they can afford.

Graduate students and postdocs are cheap labor that produces tons of science. We get what we incentivize, again and again. We've created a system that favors short-term, risk-free explorations of science, and we don't just get overly cautious, timid science; we get an entire career pipeline designed to keep young scientists churning out research as much as possible before dumping them to the curb. If a senior researcher wants to succeed, they *have* to participate in this, whether they like it or not.

Faced with the realities of the present situation and mistakenly assuming that it isn't much different than the system that produced them, tenured faculty are going to look at junior scientists and keep giving the same advice that they always have, and say it with full sincerity. The junior scientists are told to "just keep doing research!" under the guise of this providing a mechanism to ensure success, only for the rug to be pulled out from under them at the last minute (when these "young" scientists are in their midthirties) under the pretense of a meritocracy.

Supposedly, the incredible number of graduate students and never-ending postdoc positions are in place to provide a training ground for future independent researchers, with the cream eventually rising to the top. But the reality exposes the lie for what it is: all of these machinations are not about gaining experience but about gaining connections. Don't believe me? Check this out: just 20 percent of all institutions provide 80 percent of all faculty hires.[44] Do you have a letter of recommendation about your research on a fashionable topic from someone at the top of their game, and are you willing to move practically anywhere to accept a job? Don't worry, you'll get postdoc (and eventually faculty) offers. But if you don't fit that template, in any way, shape, or form, well, so long.[45]

And grad students and postdocs suffer for it. Even though there aren't enough jobs to go around, nobody does a thorough job explaining that reality to existing grad students, let alone prospectives. There isn't any real-world training done during the grad school experience. There are no résumé-building workshops. There are no invitations to meet with hiring managers at various companies. There are no job fairs. All that stops at the undergraduate level. If you're in grad school, the assumption from everybody paying your bills is that you want to stay in academia and become a professional scientist, and any mention of a life in industry is strictly verboten.

Seriously, I've heard the way people in university halls speak. When they say a graduate "went into industry," it's with the same cadence and somber tone that you would speak of the recently deceased.

Because of the lack of transparency regarding professional outcomes, and honest, open discussion (let alone encouragement) of a life outside academia, grad students and postdocs shoot themselves in the foot. They focus on writing papers instead of developing skills. They assume that journal articles actually mean something in the real world. They don't bother networking outside of the latest conference for the Meeting of the American (insert field here) Society. They assume that they will be the ones to *make it* in academia, and so they don't bother preparing for a life outside that bubble.[46]

We pretend that this system produces the best scientists humanly possible. One, I'm not even sure that would justify the enormous cost we're putting on real human lives—maybe we could afford for science to be a little less, um, *intense* in exchange for greater well-being for tens of thousands of people.

But the process doesn't even do that. The system produces people who are best at running the Gauntlet. The people who are the best at cranking out research as efficiently and as profusely as possible—and the people who just end up getting lucky. The people who succeed at the Gauntlet aren't necessarily the people we want doing long-term academic research. The people best capable of navigating a decade of inhuman stress aren't necessarily the best at investigating the nature of the physical world.

Imagine if you snapped your fingers and magically swapped this entire year's worth of academic faculty picks for their runners-up. Would anyone even notice? Would we actually be *more* productive? Who knows—we don't get to live in that universe until we create it.

Not only is the Gauntlet destroying personal lives in the pursuit of precious few faculty positions, but it's also damaging the progress of science itself by creating a brain drain. Many smart, capable, intelligent people who would make excellent scientists look at the road ahead of them and simply walk away. Or worse, they take on positions in grad school with no interest in science itself. This is especially true in data-heavy fields. Promising young students who make for excellent data scientists or machine learning experts want to end up in Silicon Valley, making tons of money solving the many technological problems of the world. They see a science PhD as just an amusing way to pass the time as a part of *their* nonscience career ambitions.

They're not in it for the discovery. They're not in it for the curiosity. They're not in it for the science. They're in it for their own reasons. They'll do an amazing job, for sure, but then they're gone, leaving the open faculty positions for, ironically enough, less-qualified candidates.

We are lying, constantly, to the next generation of scientists. We encourage people, day after day, to explore careers in science and to follow their curiosity. But we don't tell them that *we have too many scientists*. We have a surplus. Adding even more humans into the mix won't solve a single thing about this situation, let alone lead to greater understanding of the universe around us.

The current generation of scientists, the ones firmly ensconced in their offices and laboratories, are gluttons. Fighting over every scrap of grant money, squabbling over grad students and postdocs whom they can exploit for their own career advancement. They have it all—a guaranteed lifelong position, a stable income, a home in the suburbs, a career filled with exploration—and yet they want more.

And the universities themselves? They're just raking in the cash like falling autumn leaves. They couldn't care less about the futures of their undergrad and grad students, beyond the extent that positive stories help sell more tuition plans. They're not interested or invested in the long-term viability of science as a field of human endeavor. As long as the student loan money keeps on flowing, they have absolutely no incentive to change the current paradigm.

There are, as always, a few simple but difficult ways out of this mess.

One is just to let time do its work. Maybe this problem of PhD overproduction will solve itself. Young scientists are no fools after all, and with enough visibility and enough discussion even the obfuscating effects of universities and faculties won't be able to cover up the reality. It may take a decade or more for a new generation of bright young minds to realize that they are being deceived, and that the dreams of their ambitions don't stack up against the cold hard numbers of their future reality. We're already seeing these trends accelerate, with more and more graduate students leaving academia as soon as they get their advanced degree, viewing their training as nothing more than a grueling but necessary step to achieve their decidedly nonscientific dreams, leading to a further exodus of brilliant minds that could have spent decades productively answering some of the universe's most pressing questions.[47]

Or maybe undergraduates will weigh the potential benefits of a four-year degree against its absurd cost. Perhaps then undergrads will just stop wanting a career in science, because they can't afford it, with fewer students enrolling. Without the tuition money rolling in, universities won't be able to hire as many graduate students, which will ease the pressure on hiring and advancement decisions, allowing the equilibrium to return to something like it was—careers in science being difficult, but not impossible.

Maybe the coronavirus pandemic, which caused universities around the world to shut their doors and attempt a haphazard remote educational environment, will solve it for us. Maybe new rounds of high school students won't want to bother with college once the superficial trappings of university life—the fancy gyms, the nice buildings, the endless food—are stripped away, revealing the subpar educational experience for what it is.

While this is a solution, it's a painful and regrettable one. I don't want kids to become jaded. I want people to dream, to be curious, and to follow their passions. But I also want to be honest and to manage expectations. So it's better to get in front of this and try to implement solutions now to gently guide science to a healthier training and career ecosystem, before the whole thing collapses under its own weight.

The first step is honesty. Real, open, sincere honesty, from all levels of the scientific career path to every other level. The honesty needs to start with the universities. One, we need to hold institutions accountable—we must require them to track the long-term progress of their graduates, discovering what kinds of career paths they take years and decades after their receive their degree. I don't think universities will do this willingly because if the information was valuable to them (in the sense of naked greater financial gain) they would already be doing it. So force them we must, either through government action, funding agency requirements, or unrelenting pressure from parents and young students.

Those universities must make that information publicly available. We must require them to include this information in their glossy promotional mailers and be available on each department's website. At every step of the way, from undergraduate to graduate to postdoc, every single young scientist must have the information available to make an informed decision. They are, after all, the ones investing their time, their money, and sometimes their sanity in pursuit of science. We owe it to them, if for no other reason than our own sense of integrity, to allow them to make a fully informed decision.

That honesty must include the fact that postdocs do not help in careers outside of academia; that if a graduate student wants to go into industry, their best shot is immediately after graduation, not after they've run the Gauntlet for a few more years.[48]

Will all this honesty result in fewer students cramming academic halls and fewer graduate students moving into postdocs? Probably. Will it result in less science? Likely. Will it make science worse? I don't think so.

As a part of this package, we must insist that universities open up their hiring and promotion criteria. We've already met the San Francisco

Declaration on Research Assessment, which pitifully few universities have pledged to.[49] Part of that declaration includes making the hiring process visible to all, or at least pledging to avoid journal-based metrics and instead assess the actual scientific value of a candidate's research. I know that departments and committees will be incredibly resistant to this. They will argue that it's impossible to quantify exactly how one candidate is a better fit than another. They're right: it's impossible to quantify because the actual process they say they use (a meritocracy based on fair standards) differs from what they actually use (lazy filtering based on journal impacts and *h*-indices, connections, prestige, recommendations from people they know).

As an aside, if departments require undergraduate research experience to properly consider candidates, then they better start offering a lot more opportunities for those same young students to perform research. By insisting on research as a requirement, but them limiting the supply of available opportunities, they are unwittingly entrenching a system where the privileged few (ones at well-funded, top-tier universities with connections at the ready) seem on the surface more qualified than the underprivileged, but not because of any significant difference in research acumen.

If announcing the hiring criteria beforehand seems too difficult, here's an alternative challenge: explain yourself. If ten postdocs make it to the shortlist of potential faculty candidates and only one gets hired, write up a brief summary of *why* the other candidates were turned down. Scientists crave data, and we owe it to our younger colleagues to give them the information they need.

The last piece in the honesty puzzle is to acknowledge that science departments explicitly are *not* training centers for future scientists. Even if they pretend that they are (which the vast majority do), the raw fact that the vast majority of their graduates go on to nonscientific careers exposes it for the facade that it is. For example, dance departments across the country know and accept that most of their graduates will not become members of a dance company, especially for their entire careers. *And they say that up front.* No dance BFA or MFA has an delusions about the reality of the situation. Instead, the departments and the students agree that they will learn valuable skills that *may* prepare them for a career in dance but that also perfectly equip them for a multitude of careers.

If the dancers can figure this out, why can't the scientists?

If the vast majority of science trainees do not become academic research scientists, then we need to adequately prepare them for life outside academia. It's not enough to say that most graduates from most fields end up finding jobs (which is true; the unemployment rate for most science graduates, especially doctorate holders, is consistently lower than average). We have to ask ourselves this: are we serving these students the best when we don't even *talk* about life outside academia and just let them fend for themselves the moment they leave our campuses?

Many science departments and institutions are beginning to recognize this fact and are taking positive steps to train young scientists in the skills they actually need, not the skills that senior scientists want them to have. For example, the iJOBS program at Rutgers University assists biomedical students to make the jump away from academia.[50] New York University's School of Medicine has their own program, recognizing that the skill set desired by industry doesn't always line up with the skill set of future researchers.[51] The National Institutes of Health has developed career training materials for graduate students,[52] with more training available for life sciences majors from *Nature*,[53] while the Erdős Institute offers workshops and industry connections for STEM graduate students and postdocs.[54] The University of California, Berkeley, and the University of Colorado Boulder go all the way, hosting job fairs for graduate students.[55] Finally, the National Academies have called for serious reevaluation of how we conduct graduate education.[56]

To help with that program, we need faculty to have more connections to industry so they are at least somewhat aware of how the world works outside the ivory tower. We can do this by rewarding faculty for having industry experience (say, by "reverse postdocs," where future faculty are expected to spend some time away from academia) or by encouraging industry connections, even for deeply theoretical work. I can speak from personal experience that the tools and methods we develop in astronomy, from handling large datasets to sophisticated analysis methods, would be extremely valuable to many private companies—why can't those kinds of connections become a part of the scientific experience? Are we worried that we won't get enough "pure" research done? Again, the only objections I can imagine come down to a worry that we will reduce scientific output, which as we saw before isn't necessarily a bad thing.

At least universities recognize the potential monetizability of these connections, through partnerships, patentable devices and techniques, and development of new business practices. For once, the interests align (for some random examples, check out the note),[57] but we need to do a lot more work to bring these practices into all fields of science.

All of these connections to industry won't matter unless we get the faculty on board with how they engage with their advanced students. We could either ask them nicely, which probably won't work given the enormous pressures that they face to win grants and write papers, or we could make them do it. Nearly a third of recent PhD graduates have no job lined up at graduation. We need to change that. We could make job hires a requirement for faculty: In order to get tenure and win promotion within a department, they must assist their students in getting nonacademic jobs. They must ensure that they attend career training sessions and job fairs. They must acknowledge the reality that the majority of their students will not follow in their footsteps—and they must be rewarded for constructing the best opportunities for those students.

They must also be held accountable for the working conditions of their students and postdocs. Graduate students, especially advanced ones in the latter half of their training, must be treated for what they actually are: not students at all, but junior colleagues actively engaged in a research enterprise. Otherwise known as *a job*. Like any job, their rights must be protected and enforced. We need to support effort to unionize graduate students so that they can collectively bargain for better wages and better treatment. I would hate for the situation to come down to having grad students punch their time sheet when they enter and exit the office or lab, but if there is no other mechanism to enforce an adequate work/life balance, then so be it.

Yes, less science will get done, as measured by raw paper output. But the students working will have better mental health, and so it's likely that the science that does come out will be higher quality.

Lastly, we need to radically change the proportion of undergraduates to senior research positions. We can attack this from both ends of the spectrum.

On the senior end, we need to increase long-term funding support from institutions and grant agencies. We need to convince the public at large and policymakers that it's best if we *invest* in scientists, not just pay them for a few years and discard their empty husk when we're done with them. We need to create more research posts—not faculty positions but just research *jobs*. In France, where I worked for one of my postdocs, the French national research agency funded a relatively large number of permanent research posts in all fields of science. The pay was terrible compared to what those same people could get in industry, but if they had a passion for research, there were nontenure, nonfaculty routes available to them. I'm not saying that the French system isn't without its flaws, but we can learn valuable lessons from it.

And while we're at it, we need to kill the concept of adjunct teaching faculty for the embarrassing stain that it is.

On the other end, we need to reduce the supply of new scientists coming into the field every year. Some of that will come about naturally from increased transparency, increased discussion of alternative career paths, and increased training for nonscientific jobs. I'm positive that if more students at more stages had more options available to them and were encouraged, fewer of them would advance into graduate school, postdocs, and faculty appointments, reducing the crushing pressure at all those stages.

But we're also going to have to limit the supply artificially. We can place caps on the number of graduate students that departments can hire every year. We can limit the number of short-term grants awarded every year and limit the number of postdocs that any one faculty member can hire at any one time. Heck, we can even put a cap on the lifetime number of students that any faculty member can hire.

To do this, we need to limit the number of short-term grants that are awarded every year and limit the number of graduate students and postdocs that can be paid via grants.

We need fewer young scientists. I know this sounds crazy and self-destructive, but we need to try *something* to correct what science is becoming. Young scientists who want to become old scientists don't find out about their ultimate fate until two decades into their career. That's two decades that they could've spent studying other subjects, pursuing other opportunities, saving money to buy a house, having a kid, whatever.

If we're not going to give young PhDs a permanent academic research job, could we at least be a little bit merciful about it and tell them when they're still young and agile? The older you get in one career, the harder it is to switch tracks, and ultimately the less money you make in your life. The current system is designed to produce a lot of science and a lot of papers. But we're robbing young scientists of their futures to pay for the intellectual diarrhea of our faculty.

If we drastically reduced the number of graduate school and postdoc appointments, a lot of young dreams would get crushed. But dreams are getting ground into the dirt every single day anyway. We would just be doing it at a stage where we can actually be honest about it. "Most of you will not make it into grad school, and that's okay, because we're going to teach you some seriously cool stuff" is not the worst way to begin Freshman Science 101.

And maybe by closing off the pipe earlier in the career line, we can reduce the amount of toxic competitiveness. We can let kids dream about the future and, if they decide to pursue a career in science, give them the gift of an open, welcoming, trusting community.

If we don't do any of this, and we just let the system continue to burst at the seams, it will eventually destroy itself. Either we'll curse ourselves with a generation of youngsters with absolutely no trust in the scientific program to deliver careers, or we'll see science as a profession crack under its own weight, transforming into a broken, disfigured, hollow shell of what it could be.

4
The Politicization of Science

NOBODY CARES ABOUT MY WORK, and that's just fine by me. I say this only partly in jest. I know many of my colleagues wish people cared more about their research, and that their particular field had the fame, notoriety, and cool images splashed over the TV that astronomy does. When I sit next to a random stranger on a plane and they ask what I do for a living, and I say, "I'm an astrophysicist," I get a variety of reactions all centered on the locus of "generally awestruck." But I have to remind myself that their reaction isn't about me personally but my profession.

If I were to say I was an entomologist, I would probably just get a puzzled look in response. I'm sure all my entomologist friends who are reading this right now can back me up on this.

At least astrophysicists get some sort of name recognition, but beyond that . . . nobody cares. For example, at the time of this writing, there's a big debate going on in cosmology circles about the expansion rate of the universe. Is it around sixty-eight kilometers per second per megaparsec, or more like seventy-two kilometers per second per megaparsec? Believe it or not, this is a source of major controversy, with papers slung back and forth, raised voices at conferences, and entire careers put on the line.

I swear I'm not insulting your intelligence when I say you probably don't even know that those numbers or those units mean. You're a smart cookie, but you also haven't spent a decade training in this very specific subfield of science.

The result of that dispute doesn't matter to the general public. It won't affect electricity bills or traffic on the commute. It won't change tax rates or the availability of avocados in November. This allows astrophysicists like myself to toil in relative isolation, peering through our telescopes and squinting at our computer screens, free and aloof from the travails of earthly existence.

In other words, that disconnect entrenches the view that science should not involve itself in political matters. The work we do should sit above and separate from the realities of . . . well, reality. We shouldn't make it dirty, because a scientific enterprise that is free from political influence is the best kind of scientific enterprise there is: free, curiosity driven, pure.

Science should be funded because science should be funded, period. It's a worthwhile endeavor in its own right, and occasionally it comes up with something useful, like a vaccine or digital cameras, which the public enjoys.

Scientists enjoy a privileged position of support from the vast majority of the public, a position they dearly want to protect. They don't want to change the status quo, which would lead to shrinking grant funding and declining public trust and respect.

And so the vast majority of scientists are extremely nervous about wading into political debates. They don't want to be seen as partisan, or ruin their credibility, or—most importantly—make the folks in charge decide not to fund them (for discussions along these lines, check out all the references in the source in this note).[1]

It's not like this is an entirely crazy line of thinking, either. Take the ultimate cautionary tale: the story of Galileo. He was, and still is, a pretty popular, well-known, and well-regarded intellectual. He had a lot of friends and supporters throughout Renaissance Italy, up to and including the big man himself, Pope Urban VIII. Galileo got into some hot water, however, when he began to circulate some writing concerning the nature of the universe, arguing that the sun was the center of everything (aka heliocentrism) not the Earth.

Nowadays astronomy is seen as a relatively benign science, but in the early 1600s it was a seriously big hot potato. The physical structure of the universe was intimately tied to religious beliefs, and religious beliefs (especially with the growth of Protestantism) were tied to political realities. Challenging the prevailing cosmological worldview amounted to challenging the political status quo.

The pope told Galileo he could write about heliocentrism but with one caveat: Galileo had to offer a balanced perspective and not advocate for one position over the other. So Galileo went and wrote a book, *Dialogues*, in which the character supporting an Earth-centered universe was named Simplicio. That character, as the name might suggest, didn't do that good of a job defending against the new heliocentric model.

It wasn't that hard for critics of Galileo to read between the lines: he was calling Pope Urban VIII an idiot.

Whether he intended to or not (we're not exactly sure),[2] Galileo went from arguing the scientific merits of a particular case to making an ugly political battle with powerful elites—and ending up in house arrest for his efforts. I'll leave it as a homework exercise for the reader to evaluate the goals of Galileo's financial sponsor at the time, the Grand Duke of Tuscany.

Galileo wasn't alone. Scientists have seen what's happened to their colleagues over in climatology and what the fallout was once we realized that the Earth is getting warmer. Or when the cancer researchers found a link to cigarettes. Or when the biologists had to defend teaching evolution to schoolchildren in the face of fervent opposition. Nobody wants that, for science to get all political. It's too messy, and it can kill public standing and professional careers. Science is best when scientists are left alone.

And that's a mistake.

Scientists want, deep in their academic hearts, to be apolitical. To sit above and aloof. And yet they want to advocate for change. They want to be advisors and influencers. They want to make a difference. They want funding (big surprise on that one). They want science to be successful and popular. In one survey, a full 97 percent of scientists agreed that it's appropriate for them to become actively engaged in politics.[3] Ninety-seven percent! You can't get 97 percent of scientists to agree on the selection of salad dressings in the cafeteria.

And this split personality—the twin desires for scientists to be both aloof from and deeply connected to political society—is leading to the worst possible result: science that is confused, used, and abused. In a word, science is becoming politicized, turned into yet another political tool/weapon to deploy when needed, to the degradation of science itself.

It should come as no surprise that scientists think that they benefit from being seen as separate from politics. This allows them to claim academic freedom, research integrity, and a certain detachment from society. It also gives them power.

When you're seen as separate from the day-to-day machinations of the political elite, you're given a certain amount of respect and leeway. Leaders can look to you for wise counsel, for advice, for recommendations. They can ask you the hard questions, and only you are capable of answering. And if you don't care if they're an R or a D or any other letter—you are the wise sage in the back of the court, the educated neck that guides the heads of state.

And on that public side, this detachment and separateness from the political machine allows science to claim a cultured neutrality. Whoever is in power, they should fund science. Science is a noble goal of humanity, divorced from whatever the politicians are arguing about today. It's working too: in a 2019 survey, 85 percent of respondents in the general public said that they are confident that scientists act in their best interest.[4] Separation from politics ensures survivability—the cash should keep on flowing because everybody, from the lowly citizen to the holders of the highest offices, benefit from the work that scientists do.

It should also come as no surprise that this is an illusion.

Science, as we will explore, is an inherently political operation, in many obvious and nonobvious ways. Many scientists assume that they are separate from the political world and promptly step right into it, either willingly or unwillingly—while still believing that they're not being political.

Let's start with the obvious: almost all science conducted in the world in the twenty-first century is funded by government agencies.

Some of those funding agencies, like NASA or the EPA, have specific science goals and missions in mind and won't even bother funding things outside their domain. Others, like the NSF, have broad authority to "just get some science done or whatever" (I'm paraphrasing), so they fund a variety of disciplines. Some fields of science are lucky and can reach their sticky little fingers in lots of little pots of money from across the government. Some fields are more limited: if you don't get an NSF grant, you're out of work.

But ultimately, whether it's a laser-focused NASA mission or a broad-brush NSF charter, those agencies get their money from Congress. This isn't just true for the United States, of course—if you're reading this internationally, you can just swap out the random-letter acronyms for the local science-funding agencies of your choice.

The vast majority of science is seen as politically neutral or (at best) cuts across party lines, with fans and haters alike on each side of the aisle. Even some particularly charged topics, like the safety of GMOs and the use of vaccines, can be argued for or against from both Republicans and Democrats.[5]

But it's easy to see how organizations like the EPA, whose mission is decidedly divided along strict political lines, need to play an extremely careful game. Many fields of science are already politicized because the core work in their field is not only funded but performed by government agencies. A new Democratic administration steps into the White House and gains control of Congress? Or the reverse? Ask the EPA and NASA how the shifting sands in DC can drastically affect their day-to-day operations and ability to make long-term plans.

Or maybe you can't ask the EPA because sometimes their scientists are gagged.

The science performed directly by government agencies is already politicized because those organizations exist as political animals and serve at the whim of political leaders. At the end of the day, you have to make your paymasters happy. If the political powers that be are aligned with your mission, then you can expect clear skies and easy roads. If they're antagonistic, then you have to couch your findings, keep your head down, and try to survive the term.

And if you're not American, don't you sit there smugly thinking this is just a local problem.[6]

This is a precarious situation even for science that isn't directly performed by government agencies but merely funded by it. We saw how the funding agencies need to show results in order to get congressional funding year after year (and even then it hasn't been working out so well for the past couple decades). And the lawmakers are ultimately beholden to their constituents. For science topics that are generally neutral, where nobody has any strong opinions, like astronomy, we can rest assured that a general apathy will ensure at least a moderate amount of funding for time immemorial.

For hot and juicy topics, like the environment, the tactic of the funding agencies is to find enough bipartisan support to at least allow the research to continue. You may find a random Democratic senator here or Republican congressperson there who—due to the need to play to their political base—doesn't want the government to support that kind of research, but that just means you have to work a little harder to get a majority of votes on your side. You know—the usual political game.

Members of Congress aren't debating the consequences of supernova detonation mechanisms or buying TV ads denouncing the latest findings in geology. But when it comes to a few key issues, like climate change, the environment, oil drilling and fracking, energy production in general, and the funding of science—it gets personal, and it gets political.[7]

Let's turn to another obvious case: when scientists work for corporations that explicitly put profit motives above the purity and integrity of scientific research.

For example, in 1964 the surgeon general of the United States issued a report indicating a link between cigarette smoking and cancer, at a time when 42 percent of American adults regularly smoked.[8] That report signaled a significant beginning of an end—with enough media and political attention, people finally started to quit smoking because they preferred to live a life free of lung cancer (and other horrors) rather than light up.

But that report came over a decade after the evidence started seriously piling up linking smoking with lung cancer—piling up to the point that scientists had basically figured everything out and had stopped debating with each other. To be fair, some of that is because research takes time, and a half century ago the connection wasn't immediately obvious—there are a lot of other things that can also cause lung cancer, and since basically everybody was smoking cradle-to-grave, it was initially hard to tease that connection out of the data.

But some of that delay—and the delays in the following decades in making convincing public cases that smoking tobacco would put you at risk for some nasty diseases—is due to the tobacco companies themselves. Of course it was. They had money on the line and were trying to protect their industry. But aside from disinformation campaigns, lobbying efforts, and all the usual games, they also directly employed and funded networks of scientists—real scientists with actual PhDs and expertise in their fields—to conduct research with the aim of minimizing or contradicting the conclusions of earlier work.[9]

These were scientists, presumably decent, hardworking people. Presumably people who wanted to understand the way the universe works through a particular method, generating research that can't help but be biased. Of course those scientists would ignore results that didn't fit the right narrative. Of course those scientists would pursue specially targeted research lines. Of course those scientists would carefully word their conclusions to serve a predetermined narrative. Of course those scientists would appear on TV (probably wearing a white lab coat) to explain their findings and sow confusion and doubt.

Of course millions of people around the world listened to those lab-coated scientists because they wanted to trust scientists, and those people believed they're making an informed decision to continue smoking. Of course millions of people died as a result of that choice.

Or what about the scientists who work for giant oil companies like Exxon? They're doing real (and really cool) scientific research, understanding so much about the geological past and present of our planet. Without oil companies, there would be some serious gaps in our scientific knowledge. Heck, in 1977 a team of Exxon geologists figured out how to reconstruct ancient sea level changes going back almost 550 million years. It's a pretty cool plot if you ever want to look it up—it's even called the Exxon curve.[10]

But those same oil companies that employ all those geologists and hydrologists and many other -ologists are claiming that their actions are not seriously harming the environment and that the whole climate change thing is overrated—and doing a good job at steering government policy away from drastic actions that would affect their bottom line.[11]

What to do about climate and energy is a huge, thorny, multifaceted problem, and it's getting more and more political by the day. Are the geologists working for Exxon complicit in their bosses' schemes to influence policy? Perhaps they agree with Exxon et al.'s conclusion that we don't need to worry too much about the Earth's climate, in which case this isn't a big deal. Perhaps they don't like where this is going and want to do something about oil-based energy and cool down the planet, in which case you or they are a little bit perturbed.

I'm sure the scientists at Exxon sleep fine at night and can probably make a case that they can successfully do their science for the sake of science while simultaneously taking a political stance in opposition to their employers if they choose.

But to see how political and corporate processes can warp not just science but perceptions of science, we don't have to take the temperature of the Earth. We can go right to your own backyard. Take glyphosate, one of the most common herbicides in the world. Discovered in 1970, you might better know it as Roundup, and if you do, you spray it all over the place to kill those annoying weeds that keep cropping up in your flower garden.

Given how common it is, glyphosate has been the subject of intense research, especially focusing on its effectiveness in killing weeds and its lack of effectiveness in killing anything else, like your pet dog or the lining of your lungs. But beginning in 1974 and continuing for thirty years, Monsato, the developer and patent holder of glyphosate, was responsible for the vast majority of that research. And most of that research focused on its weed-killing awesomeness and less so on its impact on the environment or health.[12]

Now I will say that it seems like glyphosate is generally safe, though different regulatory and health organizations around the world have different definitions of "generally safe" and hence say different thinks about the weed killer. So we have to ask: Are all those decades of all those results about glyphosate, which has made Monsato a profit of approximately a bajillion dollars, "clean"? Can we trust the science that came out of it? Or were the scientists just paid their salaries and told what results to produce? I'm sure that the vast majority of scientists who work for Monsato or were funded by Monsato are good, honest, decent people. They might even be good scientists. But can we trust the work that they produce—even if they're right?

This kind of conflict and opaqueness surrounding biases in research infiltrates even seemingly innocuous things like the Food Pyramid (which is now called MyPlate for various bureaucrats-trying-to-be-hip reasons).[13] The Food Pyramid/Plate/Dodecahedron represents various nutritional guidelines published by government agencies, worked out by numerous committees and subcommittees. The motive is pure: we all want people to be as healthy as possible, and what you shove in your mouth has a big effect on that. The nutritional guidelines are used informally and formally, setting the scene for policymakers, health professionals, and school administrators to help us make decisions about how much of what to eat.

Are these guidelines the work of industrious, serious, honest scientists laboring over complicated studies and meta-analyses to best guide the nutritional habits of the American populace? Or are the guidelines a sneaky backdoor way for huge food corporations to ensure that their farms and products remain on store shelves? Does MyPlate represent a pinnacle of good-faith science put to use in the public's best interest (I would've gone for a pyramid pun here if they hadn't changed the name), or a cementing of corporate lobbying efforts? Do I really need to eat this much of that food group to maintain optimum health, or to maintain optimum profits?

If you're a food scientist and have accepted money from corporations in the past, are you free from bias when constructing these recommendations? Who gets to pick which research is included when making the recommendations, and why were those decisions made?

So here we have scientists, some of whom are totally and unashamedly in the pocket of Random Food Corp, and others who are just trying their darndest, trying to make recommendations for the good of the country. But they're doing it in murky political waters, and the integrity of the result ends up in question.[14]

Perhaps the most dangerous form of scientists involving themselves in politics is when they pretend they're not even doing it.

Take, for example, the March for Science, which got its start on, of all places, a Reddit forum shortly after the inauguration of Donald Trump's presidency. Trump was seen as a deeply antiscientific president, especially after he made disparaging remarks about climate change, made moves to silence/suppress/marginalize government environmental scientists at the EPA, NOAA, and other agencies, and generally made it clear that while he largely didn't care about science in most cases, when it came to very specific fields of science that were politically relevant, he had some strong opinions.[15]

Scientists, fearing an existential threat for both their livelihoods and policy issues near and dear to their hearts, galvanized into action, creating a protest movement to coincide with Earth Day on April 22, 2017. The main event was in Washington, DC, with satellite events in cities around the world.

The current slogan for the march is "Science not Silence,"[16] which immediately invites confusion. Are science and silence at odds? Do I have to pick? If I'm "silent" (politically speaking, presumably), does that mean I'm not a real scientist? Is the normal perspective of the public that scientists should be silent, and that's what they're protesting?

The first slogan, however, when the march started in 2017 was "Out of the labs and into the streets," which I remember taking offense to. I am a theoretical astrophysicist—I don't work in a lab. I work in front of a computer screen. In an office. The very march that was trying to give voice to scientists of all stripes was, right off the bat, falling into some dangerous stereotypes.

Okay, those questions aside, is that so bad? A bunch of scientists and science fans marching around talking about how awesome science is?

Well, maybe. Scientists so desperately want to be seen as apolitical, and here they were . . . being political. There's absolutely no way that anyone with two brain cells to rub together could look at something literally called a "march" and not conclude that it was political.

Don't believe that scientists are desperate to come off as apolitical? The march organizers themselves initially claimed in nice bold letters that they were "nonpartisan" and focused on policies not politics.[17] Although, curiously, that language has since been scrubbed from their website.

Perhaps the language was scrubbed because they weren't fooling anybody. The movement was clearly an anti-Trump protest, as evidenced by a later post celebrating the election of Joe Biden.[18]

There's absolutely nothing wrong with an anti-Trump protest. You don't like an administration's policies? Go right ahead and protest—it's a wonderful freedom that we should all celebrate and support. But don't pretend it's something else. Don't pretend that you're not politicizing science as you're literally politicizing science.

I'm going to claim, because I'm pretty sure I'm right, that every single person who watched the March for Science on TV immediately assumed that it was an anti-Trump event and made their judgment based on that. Maybe they supported it. Maybe they sneered at it. Maybe they didn't care. But they made their judgment on the style of the march not the substance.

If you were a fan of science already, you likely appreciated the march (and maybe even joined in—if so, good for you!). If you were not a fan of science, you equally likely did not appreciate the march (and maybe even called out against it on social media—if so, good for you!). Hearts and minds weren't won through patient dialogue—battle lines were drawn.

The main policy goals of the March for Science align mostly with environmentalism and climate change, which were the areas of science that were most at risk under the Trump administration. In the current phase of the ever-shifting sands known as the American political climate, support for environmentalism and policies that mitigate the impacts of climate change are aligned with the Democratic/liberal side of the spectrum.[19] There are other areas of science (for example, the work surrounding genetically modified organisms) that don't have strong political affiliations one way or the other, on which the march protests were silent.

The march claimed to be about specific policies, but those policies just so happened to line up with a pro-Democrat, anti-Trump (and vaguely anti-Republican) agenda. By carefully choosing which policies to highlight in the march, the organizers and participants made their intentions known: "science" belongs to the Democrats.

Let me make something clear here. Between you and me, climate change is a major issue, and we need to be better stewards of our environment. The actions that the Trump administration took to silence and suppress government scientists who were working on these critical issues were horrendous and offensive. I think we should hold politicians accountable and bring science to the forefront of the public's mind.

But if we're going to go in, we need to go all in. Why stop at environmentalism and climate change? There are so many issues and facets and angles of everyday life that science touches. By carefully curating which issues to highlight, the march organizers made their allegiances clear, which allowed spectators to tune in or tune out based on political party identity, not the issues or discussions at stake.

Did the March for Science end up increasing funding for science from all sorts of politicians in Washington? Or did it cement in peoples' minds that science—or at least, some fields of science—is a Democratic pursuit?

I was invited to speak at the March for Science satellite event in Columbus. I declined. My goal as a science communicator is to reach out, especially to communities who are normally averse to science. By pretending to be apolitical while actually being blatantly political, something called *stealth issue advocacy* and prevalent in many fields of science,[20] the March for Science made that job harder. In other words, I didn't join the march not because it was political; I didn't join because it *wasn't political enough.*

Despite the alleged aloofness of most scientists, politics worms its way into science through multiple channels. The most immediate way this happens is through the simple fact that some fields of science directly touch on and impact our everyday lives, and we as a society need to make major decisions about those impacts. And by extension, since people tend to take specific examples and broaden that to describe entire groups, when one field of science becomes political, nothing is left untouched.

For example, consider the discussion around climate change. Almost every scientist agrees that the Earth has been getting warmer at an uncomfortably fast rate for the past century, and it's mostly from us humans dumping carbon dioxide into the atmosphere like there's no tomorrow (and at this rate, there may not be).[21] That should be a relatively safe and rather mundane statement to make, causing about as much stir as the latest measurement of the expansion rate of the universe. But it's not, and the reason that even simple statements like this become contentious is because now it's not just a scientific question but a political one.

And the reason that it's a political one is that the implications of that research lead to direct, material consequences for the general public, corporate entities, and their political leaders. The problem isn't so much that the Earth is warming up—the problem is that we don't know what to do about it. Do we limit greenhouse gas emissions through a tax scheme? Do we prevent industrial development? Should everyone reduce their carbon footprint? Since different responses to climate change impact different groups in different ways (and ultimately will cost somebody, somewhere some money), the entire discussion—not just the range of policy options but the bare results themselves—get thrown into the political meat grinder.

Climate change has become the major science-motivated political battle of our generation, with politicians, the general public, and even scientists themselves accusing each other of manipulating science for their own personal or political gain.

It's a quagmire. What started as innocent-sounding scientific research on the climate blossomed into an intellectual thermonuclear war, distorting the process of science itself. Large segments of the public ask if it's possible for any climate scientist not to have a bias one way or another. And other large segments hang entire policy initiatives that affect the lives of hundreds of millions of people on the results that scientists generate. Are we getting the best scientific results possible with these kinds of life-or-death pressures?

Who's in charge here—the science or the politics? It's never an easy question to answer when the science touches on real, practical applications that have immediate consequences for our everyday lives, and that goes double for times of crisis, like the COVID-19 global pandemic. Of all the things to argue about in the midst of a global pandemic that really enjoyed killing our grandparents, we as a society ended up debating . . . masks. A simple cloth mask became a potent political symbol. Should you wear one or not? Should you be *made* to wear one or not?

Wearing masks became a *thing*. It was so much less about preventing the spread of disease and more about political identity and claims of access to reliable narratives on the nature of the disease.

And one of the biggest contributors to the reason that wearing masks became so dang controversial was that the recommendations from scientists (and especially authoritative scientists, like the ones from the Centers of Disease Control, who are tasked with, well, controlling disease in a centralized fashion) were incoherent and contradictory, especially in the early (and most critical) stages of the pandemic.

Yes, wearing masks is a good idea.

No, wait, wearing masks isn't all that essential.

No, no, we meant that wearing masks is pretty helpful, maybe.

Check that—wear masks. All the time. Seriously. We mean it.

The whole mask debate may have taken off on its own without all this, but this didn't help.[22] At every single stage the scientists making these recommendations had their reasons. At first we thought only the obviously sick would transmit the disease, but masks might help. Then we wanted to reserve the mask supply for healthcare workers. Then we learned that a certain percentage of people can carry and transmit the disease without even realizing it. Then we were trying anything at all to slow the spread.

At every stage, scientists were working with the best available evidence—and without intending to, sowing the seeds for discord and partisan war making, with deadly consequences. If we had started wearing masks, consistently, from the start of the pandemic, we certainly would've saved lives.

These two examples highlight why scientists must be so careful when they approach highly charged political issues (and why many scientists say it's better to not get involved at all). Science makes for poor politics, which makes it ripe for politicization.

First, science is slow. Really slow. The very mechanics of the scientific endeavor operate on careful investigation, controlled experiments, and detailed observations. That takes time. And that's doesn't even count the theoretical work: the mathematical development, the arguments, the teasing out data to accept or refute an idea.

It took us over a century to prove that atoms exist. Seriously, they were first proposed in a scientific context in the early 1800s, and it took decade after decade of careful experimental investigation into the world of the small, coupled with huge leaps in understanding, time and time again, to finally come to grips with atoms. It's easy to take that knowledge for granted—we learn about the nature of the atom when we're kids, but that insight was not won easily. Hundreds of scientists spent their entire careers chipping away at the problem, all for us to blithely acknowledge a reality while barely thinking about it.

Imagine if there was some major policy issue in the 1800s that critically depended on whether atoms were real or not. Imagine if major world governments had to make coordinated efforts to deal with the ramifications of the atom. What if there was real money at stake? Would the Whigs and the Democratic-Republicans have printed leaflets blasting their opponents for (not) believing in the atom? Would it become a part of presidential debates? Would it feature in bold letters in political party platforms?

All this while we were still trying to figure out if atoms existed. Can you imagine how nasty that would get, and how that intersection of science and politics would forever skew that field?

Second, scientific results always come bracketed by a list of assumptions. For a scientist, this is natural: results are always conditioned on the available evidence, and the available evidence is always limited. But policymakers need binary, yes/no decisions, not careful arguments built upon layers of nuance and always susceptible to change.

The slowness and caution of science doesn't lend itself to the expedient needs of politics. We need answers, now. There are threats and issues facing us, now. There's an election, now.

And so the politicians turn to the scientists, demanding answers. And the scientists give them answers. Provisional, temporary, guarded answers, but still answers. And the politicians get what they want—an answer—and discard the rest.

Some things we can say with a high degree of confidence (atoms exist, disease is spread by germs, the Earth is old), but in all those cases it took centuries of work in order to gain that level of assuredness. Everything else? Not so much. Science is a series of bets, where different statements are given different levels of confidence.

This makes science ripe for political exploitation. Because scientists have to express varying levels of uncertainty for any response (if they have any shred of ethics, at least), then there is always room to counter a policy or position that is "supported by the scientists." That uncertainty allows politicians and pundits to "cherry-pick" the results or papers or statements that align with their preconceived idea, because it's baked into the very tussling, argumentative, intellectual mosh pit of science itself.[23]

Yes, the Earth is getting warmer; we've had a few decades to make that knowledge pretty solid. But how bad is it going to get? How effective will various policy options be? Should I buy land in central Iowa so my grandchildren can inherit oceanfront property? Those are much, much harder questions, and the answers are much, much more uncertain—and hence the debate about global warming can drag on and on, because science doesn't provide clear answers by itself.[24]

The slowness, lack of certainty, and quarrelsome nature of science make it entirely inappropriate for crafting policy, which is why "evidence-based policy" is simply not. It's a phrase that sounds good on paper, but is frequently used by politicians when they want to invoke scientific results to bolster their predetermined platform.

It seems simple on the surface. We have an immediate social/political problem. Let's ask the scientists in some relevant field. They go out and do their science thing (whatever it is) and come back with the results. We use those results to make the best decision possible. The world is a better place. The end.

But science does not work this way, and neither does crafting policy. Science operates most quickly when it faces direct questions that can be settled with a clear and well-controlled experiment, elucidated by a sensible theory. Remember that chapter on fraud and the problem of the "low-hanging fruit" questions already solved, leaving only the nuanced, challenging, complex problems? Same thing here: most questions that modern-day scientists face simply cannot be answered with a single survey or experiment or observation.

Why are scientist allowing themselves to be used like this? Why does politicization persist? I believe it's because scientists want to both be above the political morass and be a part of it.

It's easy to blame slimy corporations, crafty politicians, outspoken special interest groups, and all the usual animals of the political circus for the politicization of science. They want money and power and will use every tool available to achieve those goals, including bending and warping science (and scientists). But it takes two to tango. In each and every case that we just looked at above, scientists themselves were involved. You can't have politicization of science without scientists participating in the process in some way. Either unwillingly (e.g., by actively getting involved in politics or by having their good intentions distorted) or willingly (e.g., with a good old-fashioned pile-o-cash), scientists are complicit in every instance of the politicization of their own field. And as we saw, it doesn't just have to be the bad apples: even well-intentioned scientists get swept up in the storm.

Scientists are among the most well-regarded and most respected individuals on the planet. When pandemics come around or when ecological change threatens our way of life, we turn to scientists for help and guidance. We lean on their expertise. We seek their wisdom. We crave their models.

And some scientists just can't resist. Either for personal glory or the firm belief that they're helping, they jump right in, offering advice and wisdom . . . sometimes with disastrous results, like what happened with the coronavirus pandemic. As soon as the disease turned from regular demic to full-blown pandemic, leaders around the world asked scientists one basic question: how bad is this going to get? In response, scientists used one of their favorite techniques for predicting the future: forward modeling.

You start with a model, which is a mathematical equation that you hope relates some things that you can know or measure (like the current number of deaths and the "spreadability" of a virus) to things that you can't know or measure (like the number of deaths at some point in the future). You also hope that your model is a faithful representation of reality—otherwise it will give you bogus, nonsense predictions.

One of the most quoted models throughout the pandemic was the one produced by the University of Washington's Institute for Health Metrics and Evaluation (IHME). It was their model that was quoted by the CDC, by the White House, by basically everybody to focus their discussions on the future of the pandemic.

It was a very popular model.

It was also very wrong.

In the early, scariest months of pandemic, in March and April 2020, the IMHE consistently undercounted the total deaths that Americans would face over the summer.[25] For example, on April 12 the IHME model predicted that on August 4 we would have zero COVID-19 deaths, indicating that by that time the disease would essentially be wiped away.

In reality there were 1,061 deaths that day.

Even as the shortcomings of the model became apparent, it was still quoted and touted and highlighted. It was scientific. It had math. It was authoritative (there were even acronyms involved!). It had power.

By the summer, whatever modeling magic they did at IHME went into overcorrect mode. In August, the IHME forecasts actually predicted more deaths from COVID-19 for the coming fall than what actually ended up happening, roughly double the number of actual daily deaths.

And yet the model continued to be quoted and cited as if it had any predictive power whatsoever.

Forward modeling and forecasting even simple systems can be a headache. Trying to read the computational tea leaves for a highly complex system as you're simultaneously learning about it in real time is a recipe for disaster.

So why did we bother? If the models were so egregiously bad, why would be even look twice at them?

Because the results provided justification for political goals, and the scientists allowed their work to become politicized. The IHME model showed that the pandemic would be bad—really bad—but then be not bad in just a few months' time.

This fit narratives on both sides of the spectrum. If you wanted to take harsh, immediate measures (full lockdowns, mask mandates, etc.), the IHME model showed how bad the spike would be. If you wanted to play things a little cooler (ignoring the disease because it's not so bad), the model showed that it would eventually drift away in the wind.

Political commentators and leaders alike bandied about the IHME predictions. It didn't matter that the rest of the scientific community had already recognized the serious shortcomings of their techniques (not helped by the fact that the IHME was frustratingly vague about exactly what they were doing and were refusing to participate in model comparison challenges)[26]—what mattered was that the IHME models were useful.

Not in predicting and evaluating the impact of the disease, mind you, but in selling a story to the public.

So no wonder some folks thought that the COVID-19 situation was overblown. Science said it was. And people died, needlessly, because of it.

Every single scientist is actually a human being, and individual human beings have their own political leanings. And it turns out, for various reasons that we don't need to get into here, that around 80 percent of practicing scientists are either outright Democrats or lean to the Democratic/liberal side of politics (the word "lean" here means that in a survey about party affiliation, they may say that they're independent, but when asked about specific policy preferences or whom they actually voted for in the latest election, they pretty much act like members of one particular party).[27] This number roughly holds for scientists in all fields, from astronomy to biology, and for academics, government employees, and the scientists working for the various corporations.

The fact that most scientists are Demotratic(ish) is fine. Even if their own personal research isn't a particularly divisive topic, they're still a member of society who wants to participate in the political world. They want to think about politics, speak about politics, debate about politics, and vote about politics. It's only natural and appropriate.

The problem comes when scientists use their scientific credentials to advance a political agenda or, worse, when they allow themselves to be used as political tools without even realizing it, while simultaneously insisting that they're impartial. This ambition creates the conditions for their message, their science, and their values to be weaponized . . . and even used against them.

Recently I've seen the slogan "Science Is Real" on yard signs, alongside other phrases like "Black Likes Matter" and "Love Is Love." I'm guessing that most scientists would not have a problem with their profession being associated with those other sentiments. I also saw a billboard advertising a pro-life group saying, "Science Is Real: Fetuses Are Alive!" I'm guessing most scientists would take issue with that.

Many people don't believe that the Earth is getting warmer, primarily for reasons of political ideology.[28] Rightly or wrongly, they believe that their way of life, their interests, or their freedoms are at risk.[29] For them, the issue isn't whether science is real or not; the issue is whether they perceive a particular policy to threaten them or not. And if science is associated with that policy, then so be it.

And on the flip side, many people do believe that the Earth is getting warmer, not because of a sound and reasonable evaluation of the evidence but because they believe that their way of life, their interests, or their freedoms are at risk. And if science is associated with policies that solve it, then so be it.

Scientists believe that they are seen as impartial observers of objective reality and thus can bring their scientific expertise to bear on a number of pressing political issues, and that because of their standing in society, their words have more weight. They're not wrong—surveys consistently show that the general public does trust scientists and what they have to say.

But scientists are increasingly using their credentials to inject themselves into policy discussions. This isn't necessarily, in isolation, a bad thing. What science learns about the world can be a powerful driving force to guide decisions.

But scientists are themselves beginning to equate particular findings or results with a political ideology. This happens in both directions, especially in the highly charged topics of the environment and climate change.[30]

When scientists reach too far and begin to associate their work with entire political ideologies rather than specific policies, it creates an atmosphere where science itself becomes a political football, something to be tossed around and argued about along with a raft of other issues, like the national debt, medical care and insurance, social security, foreign policy, and on and on. It's just one more thing that gets divided and divisive, where the people who participate in that thing and rely on it for their livelihood look to a core group of supporters in Congress—and if the political winds shift and their core group of supporters lose their levers of power, the thing risks going away.

The more that science is perceived to be a political issue, the more it becomes a political issue. Like the ebb and flow of the tide, we're setting up a situation where science funding flourishes under certain administrations and governments and flounders under others.

Simply by participating in political life while simultaneously trying to stay apart from it, scientists are politicizing science. Ultimately, as scientists become more politically active, they have to ask themselves this: how can we flourish if the people in charge don't like us?

Besides putting science itself at risk, this process of politicization results in science that gets warped, twisted, used, and abused—and scientists play right into it, time and again. And what does warped and twisted science get us? A broken relationship between scientists and the people they're trying to help.

While confidence in scientists is high (and the highest it's been in half a decade), that simple number hides some disturbing trends. For example, the political split: as a group, Republicans (or those leaning Republican) are more likely to be suspicious of scientific results and not trust the ethical integrity of scientists, believing scientists to be influenced by corruption and bias in order to favor a particular political position.[31]

In one pole, 64 percent of Democrats expressed a great deal of confidence in scientific institutions, but only a meager 34 percent of Republicans felt the same way.[32]

Republicans are even less likely than Democrats to believe in the scientific method itself: while 70 percent of Democrats agree that the scientific method produces accurate results, only 55 percent of Republicans do—meaning that the remainder of the general public believes that the scientific method can be used to produce any conclusion that the researcher wants. Fifty-four percent of those same Democrats believe that scientific experts are usually better at making science policy decisions, compared to only 34 percent of Republicans, the majority of whom believe that scientists aren't any better than average on making decisions *in their own area of expertise.*[33]

Interestingly, if you're a Democrat and highly acquainted with science, you're more likely to believe that scientists' judgments are based solely on the facts. For Republicans, it's the reverse! The more science knowledge a Republican has, the more they believe that scientists are just as susceptible to bias as the average Joe or Jane.

It's not just political lines—Blacks and Hispanics, for example, are more likely than whites to distrust scientists and think that scientific misconduct isn't being addressed adequately.[34]

Is this what we want? Do we want love of science to fall along party lines? Are we happy with accepting trust in science from no more than 50 percent of the population? Are we okay with the general public conflating scientific debates with political ones because the lines are so blurred? Do we really envision a future where a good fraction of the taxpaying public is distrustful not just of the results produced by science but of the very scientific method?

I'm going to go out on a long limb here and guess that the answer to all of the above questions is a big fat no.

But still, scientists want to engage with politics (at least, they say they do when asked in a survey), and they absolutely should. So here is my radical suggestion to get us out of this swamp of politicization: do it more.

I'm not going to sit here and pretend that I can lecture my fellow scientists, wagging my finger, pointing to all these horrible situations and outcomes that we got into because they just had to wade into the political game (and ended up getting used like any other political tool), telling them to stop trying to make their voices heard in politics. I'm not going to tell scientists that they should sit apart from politics and plug their ears anytime a policy question comes up.

And so I won't.

Scientists are going to bring themselves and their work into the political sphere because *of course they are*. They're humans. Some of them want to make a change that they believe will make the world a better place and that their knowledge and expertise can help bring about that change. Some of them want to lend their unique viewpoint to help tackle pressing issues. Some of them want to support their favorite political candidate, issue, or party, simply because they want to.

But I will tell my fellow scientists that we can't have it both ways. Scientists insist on maintaining a split personality, where they believe that due to the nature of their expertise they are both exempt from the usual political wranglings and deeply necessary to making effective policies move forward.

Scientists don't want the accountability that comes with political entanglements. They want the trappings of power and prestige but not the onus of responsibility for the consequences of that power and prestige. If you step up to the podium, you become a political target—and scientists want to avoid that. I don't blame them (I wouldn't want it either). But if we don't acknowledge this simple reality, then we will face continued divisiveness, continued segregation of support for science, and continued degradation of trust and goodwill. In the end, if we continue to allow science to be politicized, we risk damaging the very qualities that make science such a powerful political force.

Yes, there are scientists who are playing a wicked game, deliberately falsifying results (or just stretching the truth) in order to serve some narrative that they are invested in—perhaps financially. Yes, there are scientists who work for government agencies directly and are beholden to the rotating powers that be. Yes, there are scientists who take up their picket sign and argue against an administration while maintaining that they're politically neutral.

We've seen how all those choices end up with results that are far beyond what the scientists themselves realized. But all scientists, even those who shut their doors and close their ears to political issues, are complicit in these consequences every time they argue that scientists are above the political fray. They are letting politicization happen even as it gets worse.

So how about we let scientists embrace politics? After all, we can't cloister our scientists, shutting them off from society to toil in silence (even if that's what some would prefer). Indeed, studies have shown that scientists who choose to engage in policy advocacy don't risk ruining their credibility. In some cases, it can even be enhanced.[35]

Instead, we can recognize that scientists are human. They have interests and passions, desires and goals. They have beliefs (some formed rationally, some not). Like everybody else, they would prefer that the society around them mirror those beliefs. We can work to educate the general public that when a scientist speaks, it is not with as much authority as we might wish to imbue them with. In the end, the more scientists engage in politics, the more the public will view scientists as subject to bias and corrupting influences. This is unavoidable, but it's the only path forward. Scientists have to release the expectations that they can engage in politics and still be seen as neutral. Instead, they must accept that and leverage it to seek greater understanding for their particular views.[36]

In other words, it's perfectly okay to say, "Yes, I'm biased toward a particular view. In light of that, here are my recommendations. . . ." People aren't stupid. When scientists pretend to be impartial, their actions speak louder than words. At least acknowledging the reality gives scientists a *chance* to be heard.

So let's have more science in politics, but stop pretending that scientists are somehow aloof from the argument. Indeed, many scientists and scientific professional organizations are beginning to awaken to the need to bring their unique voice into the political fray,[37] pointing out specific examples of science activism success stories, like the Flint water crisis and the Volkswagen emissions scandal.[38]

There are many ways that scientists can join in political discussions. They can write policy briefs. They can attend or speak at a march.

They can call their representatives. They can sit on government-commissioned panels. They can talk to their friends and colleagues. Science has a lot to say about the world we live in, and that perspective deserves to be heard. It's dangerous for scientists to ignore politics, as that gives politics power over science. But it's also dangerous for scientists to assume that they're better than politics, especially when engaging in political discussions.

To change, though, we have to learn some hard lessons. We need to recognize that science doesn't have all the answers, and when it does it usually takes longer than we would prefer. We need to understand that all scientific results—even the ones we agree with!—probably have enough uncertainty that we can't use them to make strong claims.

We need to be comfortable with the fact that scientists aren't going to save us from ourselves, and we can't look to scientists to make the hard decisions for us.

We need to recognize when scientists are speaking outside their knowledge base. We need to look at their sources of funding and training to raise legitimate questions of bias and validity.

We need to encourage scientists to stick to their ethical guns and always insist on giving politicians, corporations, and policymakers the real (messy, ugly, uncertain) answer.

We need to understand that scientists work best when they're elucidating choices, not making decisions themselves—when we can use the tools of the scientific method to explore the consequences of various policy options, not necessarily make a recommendation.

More and more, scientists get pulled into politics—whether they want to or not. Every time they do, no matter what they say or how they do it, they become somebody's enemy.

So instead of getting drawn in like lambs to the slaughter, let's have scientists walk proudly in, heads held high, knowing exactly what they're getting into, knowing exactly what they want to say, and knowing exactly what the consequences are. Perhaps—perhaps—the more that scientists are honest with themselves and with the public, the less they'll be used as tools and weapons and the more they can build and grow trust with all citizens, not just the ones who agree with them.

5
The Disdain for Nonscientists

Is science special?

I know, weird question, but hang in with me here.

IT SEEMS LIKE A SILLY AND OBVIOUS QUESTION AT FIRST: *of course* science is special, that's what makes it science. The combination of empiricism—caring most about observations you can make in the real world—and rationality—explanations for natural phenomenon must be rooted in natural phenomenon—is pretty dang powerful. Armed with those tools, centuries of scientists (and millennia, if you really want to stretch the definition of the term "scientist") have uncovered treasure trove after treasure trove of beautiful knowledge about the world we live in.

The universe revealed by science is both striking in its simplicity and awe inducing in its complexity. We have exposed natural laws that drive everything from the expansion of the universe to the inner workings of the cell. We have conquered diseases and mastered flight. We have peered into the furthest reaches of the cosmos and become creators and destroyers of the very forces of nature.

I'm not just saying this because I'm kind of a fan of science, but we *should* be awestruck by the power of the scientific method to discover knowledge about the universe. Science has transformed human civilization from the ground up—with our ever-more-powerful insights into the inner workings of nature, we have developed technologies that have made many aspects of modern society completely unrecognizable to our ancestors.

As a simple and immediate example, I'm typing this book on a device that, in order to exist, relies on an understanding of quantum mechanics, the physics of the subatomic world. Without that understanding, the semiconductors that drive modern electronics simply wouldn't exist.

It's so easy to extol the virtues of science by pointing to the wealth of knowledge and technology gained by scientific inquiry. And it's not just the accidental by-products of the method but the method itself that rightly deserves praise.

The scientific worldview is powerful. Thinking scientifically allows one to examine and study the world in a way that truly is unique. The scientific method is the whetstone to sharpen our intellect. It's a lens that brings the world around us into focus. It cuts out unnecessary information, reduces and eliminates bias, and focuses with laser precision on delivering knowledge. Science enables us to see the world in a clear, direct, powerful way.

That makes the way that scientists think separate from the way nonscientists think. Many modes of inquiry and thought rely on argumentation, discussion, rational analysis, and so on. But science takes it to another level. In science, nature is the ultimate arbiter of any debate, the ultimate referee of any dispute. Arguments can't just be crafted out of thin air—they need to make testable, reliable predictions that can then be evaluated against experiment. If the evidence doesn't support a hypothesis, it gets rejected like moldy leftovers.

So scientists argue with each other, just like every other community of people on the planet, but they've all agreed in advance on the ground rules for settling disputes, and so the arguments take on a different tone. Scientists rarely engage in the kind of political grandstanding or smear campaigns that you might see elsewhere—they focus on the data and the evidence. Science is messy, difficult, quarrelsome . . . and it delivers.

In short: *yay science!*

I believe that science is a good thing. It leads to an increased understanding of the natural world. It enables powerful technology. It's really fun to do, and it's capable of scratching the natural human inch to be insatiably curious in a very satisfying way.

And just like a delicious cake or an extravagant cheese, too much of a good thing can be bad.

There's this word that's been tossed around for almost a century now, and it's a word I'm going to introduce to you now: "scientism." The word itself has had—and continues to have—many varied and usually contradictory meanings, which is good for us: it means we get to assign the definition right here, right now, and explore the consequences of that.

The definition of scientism that we're going to explore is when science—a method and practice for acquiring gainful knowledge of the natural world—gets promoted to an ideology. A belief system. A value system. An engine to drive all human endeavors. An article of faith. When the only thing you need to help guide your thinking is science and science alone. When all other modes of inquiry and thought aren't just inferior—they're useless. When if you're not thinking like a scientist, then you're not even thinking at all.

When science is used in places where it doesn't really apply. When scientists make claims of a solution far outside their expertise—and the data. When scientists disparage religion, philosophy, the humanities . . . really anything but science.

In other words, scientism takes science from "good," skips right over "better," and barrels straight on to "best."[1] As we will explore in this chapter, I believe that the root cause of scientism is a pervasive belief that science itself is under attack and threatened every day by forces who wish to destroy the very institution itself. And so scientists fight back. Some scientists do that by insisting that science is The Way—not only the best way to understand the world but the best way to think. The *only* way to think. But we're going to see how scientists are fighting a war that doesn't really exist, and in the end hurting only themselves.

Scientism manifests itself in many different ways. Let's start with the first, which is good old-fashioned snobbery.

And I'll be the first to admit: for a very long time, I was a science snob. One of the main ways that scientists effect their culture of superiority—their scientism—is through the use of jargon. Jargon plays an absolutely critical role in any sort of technical field: when you're talking among your peers, you need efficient modes of communication, and so you develop special words, phrases, and conventions to make your life easier.

I can't even count the number of jargon words I either know or have encountered in my career. "Black holes," "adiabatic," "tensor field" are just three random ones I thought of right now. To you, those words might or might not be meaningless. To a physicist, those phrases are incredibly dense, offering a shorthand identifier for a very complex set of ideas.

Jargon is useful indeed, but it can be abused. Jargon can be used at a gatekeeper, a filter: if you're not speaking the right language, then you're not one of us. The overuse of jargon (which is not limited to the sciences) creates a culture with identifiable in-group and out-group codes. It's how one scientist can pick one another out of a crowd—if you use the special words, then the right people will respond in the right way.[2]

Jargon can be used as a barrier to separate the scientific elite from the masses. How many times have we seen TV shows or newspaper articles mocking people for not knowing supposedly basic scientific facts? The assumption is that if you don't know some random bit of science trivia that has absolutely no impact on your daily life, then you're, for all intents and purposes, illiterate. Jargon and scientific knowledge become tools not for easing communication about the wonders of nature but for dividing society into the intelligent haves and the unwashed have-nots.[3]

Moving beyond mere use of specialized words, some scientists take their snobbery to a whole new level, claiming that science has exclusive access to the world of knowledge—they the only way to correctly view the universe is through a purely scientific lens. As an example of this, perhaps the single most confusing statement that scientists can make is when they belittle, mock, and ridicule philosophy, calling it useless, anachronistic, or a waste of time.

Think I'm making things up?

Here's celebrated astrophysicist Neil deGrasse Tyson on the subject of philosophy, when asked about it during a podcast interview:

> If you are distracted by your questions so that you can't move forward, you are not being a productive contributor to our understanding of the natural world. And so the scientist knows when the question "what is the sound of one hand clapping?" is a pointless delay in our progress.[4]

There it is folks: philosophers are not productive. The goal of acquiring knowing is "progress" (though we're not exactly sure what we're progressing to), and anything that does not satisfy that objective is pointless. Not enough for you? During a talk, famed astrophysicist Stephen Hawking claimed that "philosophy is dead."[5]

Still not satisfied? Famed science popularizer Bill Nye said, in a video discussing whether science had all the answers, that "philosophy is important for a while . . . but you can start arguing in a circle."[6]

What the heck is going on here? It's not just limited to celebrities—many scientists known and unknown are simply forgetting about philosophy, sending the entire millennia-spanning human initiative to the trash.[7]

The general attitude that scientists have about philosophy is, as you can guess from the quotes I chose, that it's useless. That is all amounts to a bunch of pointless questions that don't have answers, with philosophers standing around arguing about meaningless concepts of no utility.

That's the greatest sin that one can commit in the scientism worldview: lack of utility. If you're not advancing knowledge, then you're not worth doing. No, wait, let me refine that so that it more accurately reflects the scientism position: we are first going to define what "advancing knowledge" means, and it just so happens to line up with the way we were doing it already, so we can proclaim in a very nonscientific way that what philosophers are doing is lame and what we are doing is awesome. Case closed.

Are scientists forgetting who they . . . are? The letters that I get to type after my name are *P*, *h*, and *D*, which stand for . . . wait for it . . . Philosophy Doctor. Doctor of Philosophy. As in, different than a medical doctor or juris doctor. A doctor of philosophy. That's not an accident or a mistake.

I'm a literal philosopher.

Scientists are philosophers. Remember a few pages ago where I extolled the virtues of science, of its powerful two-hit combo of empiricism and rationality? *Just where do you think we got the ideas of empiricism and rationality?* We didn't find those concepts on the top of a mountain in the Alps, folks; those are philosophical concepts.

The old name for "scientist" is "natural philosopher," and that term held sway until the mid-1800s, when the word "scientist" was deliberately coined to designate a profession as a sort of counterpoint to "artist." But just because the labels changed, it doesn't mean that the underlying truth went away. There's a reason that the degree and title retain that heritage.

Science is a branch of philosophy. It does not exist separate or apart from philosophy. Philosophy gave birth to science and continues to inform it.

Take the concept of falsifiability, a core tenet of modern science. In science, we now believe that we should only make statements that can be testable—proven false with some experiment or evidence. This idea is wholly ingrained in the modern scientific makeup—it's simply a matter of our regular training nowadays. And guess where it came from? Karl Popper . . . *a philosopher.*

I'm more than willing to forgive someone for thinking that philosophy is dead or useless, because chances are that they're not a working philosopher under a different name. Scientists have no such excuse (although to give credit where credit is due, after an outpouring of criticism, Bill Nye did end up admitting the value of philosophy).[8]

The arguments against philosophy represent a sort of tunnel vision that scientists can get, where if it isn't their specific way of examining the world, then it's somehow not contributing to the advancement of human knowledge—and guess what, *they get to define what "advancement" means.*[9]

Scientists charge philosophers with the crime of asking the same questions for generation after generation, ignoring the fact that they're doing the exact same thing: in my own field, we've been asking "How old is the universe?" since basically forever. Neil deGrasse Tyson publicly bemoaned philosophy because "it devolves into a discussion of the definition of words" . . . while simultaneously arguing that we should change the definition of the word "planet."

I'll put it simply: scientists do not have a monopoly on knowledge. They don't even have a monopoly on empiricism. Science is not the only path.

Do we really expect scientists to have all the answers?

I'm asking that rhetorically because you already know that my answer to that question is no, but champions of scientism believe that the answer is a firm and hearty yes, as ridiculous as it sounds. This mindset completely ignores the vast array of interesting, useful, powerful questions that philosophers across humanity have asked for thousands of years. Does science really have the path to an ultimate answer for questions about existence, values, the nature of the mind? Can science provide a morality for us? Explain the very nature of reason itself?

I'm not asking if science can *inform* those questions—it most certainly can and does, to the delight of everyone, philosophers included. But scientism takes it one step further (as it always does) and claims that science is *the* way to answer those questions, and many more beyond philosophy itself. As in, some proponents of scientism advance the idea that science can explain literally everything.

No, I'm serious. Some scientists (and science advocates), like Sam Harris and Steven Pinker, assert that the work of philosophers, historians, lawyers, artists, and anything else belonging to the general class of "the humanities" should more properly be viewed as a branch of science.[10]

If I were making this up, I don't think you would even believe me. I know I wouldn't.

Need help defining a moral compass? Believers in scientism say that science can provide a way. Trying to understand the inner working of an ancient culture? That's best viewed through a scientific lens, thank you very much. Wanting to concoct a just set of laws to govern a country? Don't worry—science has your back.

Again, science can inform all of these questions. There's always a place for empiricism, rationality, falsifiability, and induction at the table of intellectualism. But it is not the whole table, and it should never be confused as such.

For example, if you claim that the scientific approach is the best way to solve some long-standing question, how did you arrive at that statement? Was it through the formation of a testable hypothesis and the scrutiny of observational data? Or did you just . . . say it? You know, in a very nonscientific way? A full application of scientism reveals the farce that it is: if you're going to claim that science has all the answers, then you can't just assert it. If you do, then you're not even following your own triumphal advice.

Scientism has led to the use of science in areas of human activity that do not easily resolve themselves into falsifiable, testable, data-driven chunks. For example, science can tell us about the psychological and sociological nature of people, but are we supposed to trust science to tell us who to hire for a job?[11] Science can deliver gene modification technologies to increase crop yields, but it struggles on how to convince people that it's safe.[12] Through science we can have a better understanding of human health, but can it devise a nationwide nutrition program?[13] Science made opioid painkillers possible—relieving an innumerable amount of suffering—but doesn't know how to grapple with the complex opioid crisis ravaging rural communities.[14] Through science, we know that the Earth is getting warmer, but it fails at explaining why many people have a hard time believing that.[15]

These are all a result of scientism—the belief that not only can science deliver exciting new information about the universe but it can also guide human thoughts and feelings. That science can command society, not the other way around. That if you have a question, science has an answer.

The truth is, folks, that many systems—especially those involving us wonderfully complex humans—can't be studied in the way that science prefers. There are too many variables, all intertwined, and not easily testable.

How do we create a fair and equitable society through the writing of a constitution as a basis of government? Is a constitution even the best way to form a government? Are you (not *you* personally, I'm speaking to an imaginary advocate of scientism here) seriously going to argue that the only way to answer those questions is through science?

Here's another one. It sounds cheesy, but what is love? Yes, we can certainly provide *an* answer to that question through the lens of science. And that answer would probably involve a lot of chemicals. That answer isn't wrong (it's been tested!), but it's also not complete. There is *so much more* to the sensation of love that is not captured by a scientific answer that it's laughable to argue otherwise.

Here's one that's not so cheesy: what is time? Sure, science has a lot to say on the nature of time, involving statements about relativity and entropy. And it also has a lot of ignorance. We don't know, scientifically, what time is, and yet it's something that we're all intimately familiar with. That's why when I wanted to create a public outreach performance exploring the nature of time, I worked with a modern dance company; the dancers and choreographers had just as much to say about the subject as I did, if not more so.

And it's not just a matter of science not having had enough time to generate satisfactory answers to those difficult questions. One, exactly how long are we expected to wait before science can provide an answer, and what are we supposed to do with ourselves in the meantime? And two, it's simply not a matter of having enough time. Claiming that science will eventually have all the answers is a highly nonscientific statement to make. At *best*, it's a hypothesis, and not much of a falsifiable one since you can always ask for more time.

A belief that scientism is correct—that science is the superior and sometimes sole form of acquiring new knowledge—is just that: *a belief*. You can believe it all you want, but don't expect anyone else to get onboard that train with you. But it still doesn't stop many scientists from promoting it.

It doesn't help matters that many people respect and admire scientists and generally expect them to have the answers. For example, you wouldn't believe the number of times people ask me about God. Do I believe in God? Do I think God exists? The answer to the first question is "None of your business." The answer to the second question is "Why the heck are you asking *me*?" I'm a scientist, and what does my training or expertise as a scientist have to do with the existence of God, let alone any form of deity? Sure, many scientists are atheists, but that tells you more about the kinds of people who end up doing science than anything empirical about the nature of the divine.

If you're looking to science to answer matters of faith, then you're barking up the wrong tree. If you want science to provide you with an answer to questions of human faith, *then you already have the answer you want.* What people believe in and why they believe can be studied with science but not answered with it. What do I know of the existence of God, outside of my own (possibly non)belief? I've studied the Big Bang, the history of the universe, the contents of the cosmos . . . and yet nowhere in my studies did I find anything resembling a solution to *that* question. So again, why are you asking me, as if I had some sort of useful answer? I could tell you what I personally believe, but I won't, and so there's nothing more to say.

But many scientists do have something to say and argue that science *can* (dis)prove the (non)existence of divinity. Are we supposed to take them seriously and listen to them every Sunday at the lecture podium instead of the pulpit?

Why are scientists doing this? Why does scientism flourish? Why aren't scientists happy being scientists—respected, admired members of the society and contributors to the flourishing of human knowledge? Why isn't this enough?

I mentioned it earlier in the chapter, and now it's time to dig into it more deeply. Many scientists believe that they are under assault. They believe that science as an institution faces many enemies on multiple fronts. They believe that if they don't claim new ground for themselves, they put themselves at risk of losing it all. Many factions are out to get science. And scientists need to strike back.

Is science really persecuted?

I suppose everybody and everything needs an origin story nowadays. It's not just good enough to have your favorite superhero running around smashing bad guys; no, we need to understand their tragic past that motivates their need to smash.

And science has one heck of a tragic backstory, at least as the traditional narrative goes. And the primary villain in this story is none other than religion itself.

We've all heard the story, so I'll repeat it briefly here. Early philosophers, mathematicians, and theologians were starting to make interesting observations about the heavens in the late 1500s. These observations, mostly concerning whether the Earth or the Sun was at the center of the universe, bucked the established thinking at the time. These protoscientists started to write about their findings, and the way they argued was revolutionary: they were grounding their discussions in evidence and observation, not theology or dogma.

Naturally, the Church-with-a-capital-C (as in, Catholic) had a problem with this because the evidence they were uncovering endangered their earthly power, not to mention their narrow-minded faith-based views of the cosmos. The protoscientists were threatening Catholic orthodoxy and teaching in a time of rising Protestantism, and so those protoscientists were excommunicated, placed under house arrest, and burned as heretics. You know, the usual horrible Middle Ages stuff.

But evidence is evidence and science marched on, and centuries of struggle brought us to our current, modern, sophisticated, *rational* view of the universe. But still, remnants of that struggle remain, and the church (not so much the capital-C kind, but various religious denominations across the world) continues to fight against the ever-growing tide of reason, rationality, evidence, and science.

This is called the "conflict thesis," which was first proposed back in the 1800s, right around the same time that "scientist" became an official job title, as a way to explain the origins of science and its relationship to religion: one of a contest between worldviews that maintained radically different approaches to understanding the universe and our place in it. It was and is a turf war, a zero-sum game, where the winner gets to guide humanity in the direction it pleases, and the loser becomes a footnote in the pages of history, a mistake on our glorious march of progress.[16]

At first glance, it's a very attractive story, and there's plenty of examples to cite in favor of it. For starters, Galileo was indeed placed under house arrest for insisting that the Earth orbited the Sun. To us, that seems rightly heinous, a man charged with the crime of being scientific in an age of superstition.

Or there's the story of Giordano Bruno, as told by the popular TV show *Cosmos* (the new one with Neil deGrasse Tyson, not the old one with Sagan). According to the show, Bruno was a poor friar who was burned at the stake by the Roman Inquisition for the crime of hypothesizing the existence of worlds orbiting other stars.

Even in the present day, hundreds of years after that whole burning-heretics thing, we can still find evidence for conflict between science and religion. We have the Scopes trial in the 1920s in Tennessee to decide if evolution should be taught in public schools, with resurgences of the debate popping up every few decades. We have the heated issue of women's right to abortion, with scientific perspectives clashing head-on with religious beliefs. We have the Catholic Church's stance on birth control, and many religions' oppositions to genetically altering humans and using stem cells.

Many eminent and popular scientists, like the late Stephen Hawking, have espoused support for the conflict thesis, with Hawking going as far as to predict the ultimate victor in this great intellectual struggle, saying "science will win."[17]

The scientism as expressed by the conflict thesis can create absurd depictions, like the popular Christian "fish" symbol given legs with "Darwin" written inside. I once saw a three-paneled portrait of Copernicus, Galileo, and Kepler in the office of a CEO of a major science education institution. When I first saw it, I was very confused because it was arranged exactly like a triptych—an old Christian style of painting—and it seemed odd and out of place. Does science really have patron saints now?

Astronomer Carl Sagan was one of the biggest science popularizers of the modern age, and his international bestselling book *The Demon-Haunted World: Science as a Candle in the Dark* is a clear and direct defense of the conflict thesis.[18] Most of the book describes the power of the scientific method, along with the wonder and beauty of the universe that is revealed by that method. And in those passages the book is beautiful. But Sagan can't help but also claim the superiority of science over other modes of analysis—in Sagan's view, science isn't just good, it's the best.

The title itself reveals Sagan's view of the place of religion. Is science literally the only way to fill the void of ignorance? If I had presented to you a title with any other word in the place of "science," as in _____ *as a Candle in the Dark*, would you have assumed it was a religious text?

Softer, sneakier versions of the conflict thesis abound in popular presentations of science.

Take Steven Jay Gould, the famed evolutionary biologist who proposed that science and religion were "non-overlapping magisteria," or NOMA (because if it's proposed by a scientist, it *must* have an acronym).[19] This is the view that science and religion aren't really in conflict because they focus on different domains of the human experience. While on the surface this seems like a neat and tidy solution, it's really just the conflict thesis in disguise. To make this idea work, you have to believe in one of two options. Option 1: You really don't believe that science and religion can ever have competing statements about the same subject matter, and you just ignore anything that anybody ever says about, say, the origins of humanity or the cosmos or what happens when we die. Or you have Option 2: When science and articles of faith do make competing statements, you simply pick a winner (I'm guessing that fans of NOMA choose Team Science every time). And so NOMA isn't very nonoverlapping at all, and it's really just a very diplomatic version of scientism.

While most scientists may not believe in a major conflict between science and religion,[20] they still buy into the fundamental science origin story born out of the shadow of religious and superstitious thinking. When my own education touched on the subject of the history of science, it was certainly portrayed in that way. That said, many scientists do believe that religion has its place, so to speak—and that place is very far away from anything that science can touch. In other words, they believe that there's no major conflict between science and religion *because science won.* And yet they still seem to think that science is in a precarious position.

This is a very disturbing narrative. Scientism makes it appear as if there is a conflict between science and religion and promotes it as such.[21] To the advocate of scientism, other ways of viewing the universe and our relationship to it aren't just inferior, they are antagonistic. It's not just limited to philosophy or the humanities but excludes *all* alternative ways of thinking. And this is a zero-sum game: the more religion wins, the more science loses.

If it weren't for all those pesky religious authorities in the past few centuries, the proponent of scientism exclaims, imagine how much progress we would've made!

The problem is that the conflict thesis is, basically, wrong. The history of science is not at all marked by a prolonged war with religion, or really any other ideology.[22] The early protoscientists were just *thinkers*, using every tool available to them—observations, rational arguments, mathematics, theology, historical precedent, and the ever popular just making things up on the spot—to advance their lines of thinking.

Consider Copernicus, who was the first major player to advocate for a Sun-centered universe. One of his main lines of argument was that just like Jesus is the light of the world and is at the center of the Christian faith, then so too should the Sun be at the center of the universe because it's also the light of the world.

Or Johannes Kepler, who discovered that planets move in ellipses in their orbits. Sure, he used a lot of raw observational data and some serious number crunching to get that result, but he did so because he believed that the orderliness of the heavens (the "music of the spheres" as he called it) could reveal divine intent for our lives on Earth. Ultimately, his model of the universe would become popular not for any major scientific reasons or the weight of evidence but because it made calculations of horoscopes easier (indeed, there was little distinction between astronomy and astrology, or chemistry and alchemy, until relatively recently).

Yes, Galileo did suffer house arrest for crimes of heresy against the Catholic Church. But was he punished for claiming that the Earth moved about the Sun, or because he called the Pope—his sponsor and generally the boss around those parts—an idiot?[23]

Bruno burned at the stake. But the Church didn't care about his ideas of worlds orbiting other stars—they cared about the fact that he was preaching against core tenets of the Catholic faith. You know, really minor stuff like *the existence of the soul.*[24]

Should Bruno have been burned and Galileo gotten locked in his home? Certainly not—it's perfectly reasonable to modern eyes to condemn those actions as inhumane. But was it about religion keeping a lid on science? No. Was it about religion fighting a war to maintain its turf? No again. Was it about a complex, ever-shifting set of values and worldviews evolving with time in a messy, complicated, politically charged way? Bingo.

Were we really "in the dark" before the development of modern science? Yes, we've learned a great deal since the days of Galileo and company. But those luminaries weren't separate from their respective cultures either; they were as immersed in all the thoughts and arguments and discussions as anybody else. And they didn't start from scratch. They came from a long intellectual line that stretched back generations. That line was full of thinkers and curiosity seekers who sought to understand the world around them and put that understanding to good use.[25]

Historians of science have largely dismissed the conflict thesis, for good reason. Science (and scientists) have been supported by religious institutions, government institutions, and religious institutions masquerading as government institutions ever since the scientific revolution (if that's even an accurate phrase anymore) four hundred years ago.

Even in modern times, the pieces of evidence for a conflict are few and far between. Except for a few high-profile, hot-button issues, members of almost all religious denominations are just as likely to seek scientific answers for pressing personal and social issues as nonreligious groups,[26] and the majority of people, religious or not, would be happy if their children grew up to be scientists.[27]

And yet the narrative persists. Scientists still believe that science and religion were in conflict in the past, and that when it comes to certain issues, like evolution, religion is trying to muscle in on science's turf. The majority of the American public believes it too. In one survey, 59 percent of respondents believed that science and religion were in conflict, with that number curiously jumping to 73 percent among believers who seldom or rarely go to religious services.[28]

What we see is a deep divide between the narrative and the reality. Not the first time this has happened in human history, I suppose, but it severely undermines the trust that the public has in science and the trust that scientists have in the public.

Scientists still find ways to perceive themselves to be under threat, and the public goes along with that story—otherwise scientism simply wouldn't exist. Yes, there are some factions of religions that want to rid the world of science in all its forms, but the vast majority of people in the vast majority of situations simply ignore the extremists and go on with their lives. But despite the general lack of care when it comes to this supposed Science vs. Religion ultimate fighting championship, scientism still claims that the core values of science are under threat.

And it's not just religion that threatens science; it's all of humanity.

No, wait, I mean *the humanities.*

Is science going away?

Many scientists believe that it is. And they're not exactly unjustified in believing so, as we saw earlier with the decades-long slow decline of federal grant funding. There's less money to go around these days for our fun science experiments, so I suppose it's natural to assume that it's the Enemies of Science who are behind it all.

Besides religious fundamentalists and their strident antiscience views (well, in certain limited cases, like evolution; issues like global warming and antivaccination cut straight across faith-based lines), the other antagonists to the great scientific enterprise of our modern world, at least according to fans of scientism, is none other than their academic colleagues. The people they have to see on the quad. The folks in that building over there.

The humanities professors.

"The humanities" is an incredibly broad category, encompassing some fields of study that we've already met, like philosophy and theology, and some we haven't, like history, archaeology, anthropology, and law. Taken together, the humanities encompass most of what we used to call the liberating arts (which is now shortened to "liberal arts")—the essential education that a free person needs in order to be a productive, thinking, *leading* member of society.

Science in its various incarnations, starting with its progenitor natural philosophy, was once included in these liberating arts. If you were interested in studying nature, then it was seen as much more akin to studying history or rhetoric or theology: something that enlarged your worldview, broadened your horizons, and deepened your intellect. It was decidedly different than *training* for a particular trade, like plumber or doctor or banker. It was something that only the truly elite in society could strive for (sorry, doctors . . . and plumbers . . . and bankers).

But in the 1800s, starting at the University of Berlin and quickly spreading around the world, attitudes began to shift. Science was seen more as a profession, a vocation, than one of many and varied ways to approach inquiry into the world around us. It became a job. Something to focus on. It became . . . science. And so it separated from the liberal arts. You attended a university to become either a scientist or an artist, but not both.

Fast-forward 150 years, and we have the modern university, which isn't so much universal and more of a collection of siloed departments, each pursing their own programs for their own purposes and own advancement, with little to no cross-pollination—let alone discussion—among the various faculty.

It's to the ridiculous point that scientists adopt an absurd level of DIY-ism, refusing to send their students to computer science departments to learn programming or communication departments to learn writing, even though they fully expect those same students to spend the vast majority of their time writing either computer code or a human language.

And did I mention that all these departments are in competition with each other? When students attend university, they don't come for a universal education—they pick a major. They choose. And when they choose, their precious tuition dollars flow to that department to the exclusion of others.

I don't want you to think of a university as a fertile ground for exploring all the intellectual pursuits that humanity has to offer. I want you to think of it as a collection of medieval fiefs, led by barons and dukes, pledging a vague sort of fealty to a ruler (aka the president) but generally hating each other.

So we have a state of affairs where scientists, as represented by the various faculty members in universities nationwide, don't just tolerate the humanities but *despise* them, because every student that doesn't choose a STEM career represents tens of thousands of dollars of lost potential revenue. In a world where grant money and government support don't really keep the train rolling anymore, the students are the pawns in these petty wars.

In universities across the world, the sciences and the arts are locked in a never-ending culture war.[29]

The *perception* among scientists is that they're losing funding (because grant dollars are drying up like a tide leaving a shore), and they are looking for any sort of proximal cause other than the fact that it's largely the fault of scientists themselves (as we saw), so they look to blame somebody else. And who else but those faculty members teaching "useless" things to students? Hence the claims that everything should be absorbed into the blob that is science, and if it doesn't advance human knowledge in the exact same way that science already does, then it should be discarded as a superfluous waste of resources.

And if a humanities department or two either disappears or gets folded into a STEM major, then the money attached to every student can go to the *right* place.

As we've seen already a few times when it comes to the scientism picture, the narrative doesn't align with reality. In truth, the humanities are in crisis. For one, departments across the country are seeing declining enrollment numbers, refusal of university administration to replace retiring faculty, and sometimes outright closures.[30] From the late twentieth century to the present day, venerable—and liberating!—majors such as history, philosophy, English, and literature have seen 30 to 50 percent drops in enrollment numbers.[31]

And in an interesting flip of the narrative, it's the humanities majors—who by all objective evidence are steadily losing money, students, and literal ground in campuses worldwide—who see science, and scientism, as their primary threat. They think that scientists are out to either stamp them from existence or assimilate them into their own programs.[32]

And as we've seen in some of the choice quotes I've pulled and references that I've cited, *they're not wrong.*

Ever since the 2008 financial crash, there's been a consistent and clear message to encourage students to go into STEM majors. The thinking goes that humanities degrees are "useless"—replete with jokes about philosophers flipping burgers—and so if you want to make a living in a cold, hard, postcrash world, it's either STEM or nothing.

And so we see the gold rush of students flooding university STEM programs, and all the problems we looked at early. Most notably the lack of jobs.

After a decade of poking philosophy departments in the eye and laughing at them for not having enough jobs to support their numbers of students, scientism did the exact same thing to the field of study that they claim is the best.

Gee, if there were only some sort of systematic method of combining observations and rational explanations to understand the relationship between cause and effect.

What do we do about it?

Like any multiheaded hydra used as a metaphor, scientism can take many forms.

It can be the simple use of jargon to exclude others from the community.

It can be an insistence that philosophy, of which science is a branch, is useless. Or boring. Or stupid.

It can be an assertion that other fields of exploring and understanding the human condition here on earth, like history and archaeology, should really just be branches of science . . . or tossed out because if it's not science then why bother?

It can be a claim that science is the sole arbiter of truth and knowledge, lone provider of value and morals, and that all questions that could possibly be posed are either (a) answerable with science, given enough time and resources, or (b) not real questions.

It can be the embracing of a false narrative that religion is out to stamp out science. And that somehow they're working in concert with the disciplines of the humanities.

It can be a lot of things, done or expressed in some form or another by almost every member of the scientific community.

And it's one of the many things that strains and ultimately will break the relationship between science and the public. If scientists continue to insist that they're better than everyone, smarter than everyone, cleverer than everyone, better looking than everyone (okay, okay, that one's a stretch), then why should they be surprised if everyone else turns away from science in disgust?

But nobody likes a self-described elite class, which is exactly what scientism claims science is.

Every time a scientist claims they can answer a question that's been bedeviling another field for ages . . . are they really convincing anyone? Or, on the other hand, are they convincing everyone that the scientists don't know what they're talking about and have very greedy fingers?

Every time a scientist looks down on another discipline or mocks *an entire field of human inquiry*, how are people supposed to respond? With open arms and a willingness to be told exactly how wrong or useless they are? Or let out an exasperated sigh?

Every time a scientist claims that they're under attack from religion, do they find support and sympathy from the general public, or are they viewed as just another muckraking partisan?

Every time scientists make themselves inaccessible, unassailable, and infallible . . . do they just look like fools?

All that scientists are accomplishing by pursuing the narrative of scientism is building walls between themselves and the public . . . the very public they rely on for their funding support. Don't bite the hand that feeds you, right?

I see the existence of scientism as a reaction by scientists to imaginary—or at the very least perceived—threats. They believe that religion tried to stamp out science in its infancy, killing it in the crib before it could even walk, but that only the dogged determination of science has allowed it to survive all these centuries—and that science continues to face faith-based challenges to the present day. Or that once science established itself as separate from the liberal arts, the humanities have been trying to gun down their academic rival ever since.

Fueled by the specter of those supposed menaces, scientists seek revenge . . . and scientism arises.

Neither of these threats are true, and yet the stories persist. I honestly can't give you an explanation for *why* those narratives remain despite the evidence. You wouldn't think that scientists of all people, who are best equipped to root out bias like the stubborn weed it is, would be susceptible to this. But then, despite what champions of scientism claim, scientists are just normal people too.

So what do we do about it?

Let's start off with a radical suggestion: let's celebrate the fact that science has limitations, and that *limitations are a good thing.* Science is so amazingly powerful *because* it focuses so well on questions that it has the capacity to answer. You know that phrase, if all you have is a hammer, then every problem looks like a nail? Well science is a pretty awesome hammer, and there are a lot of nails in the universe. Yes, the intellectual precursors of science run back millennia, but something magical did start happening about four hundred years ago that really got us moving, with knowledge of the natural world growing exponentially. Obviously science is doing something right, because I'm pretty sure that we understand just a little but more about the world today than we did yesterday. It's a method that knows how to deliver results.

But science isn't equipped to answer with complete satisfaction every single meaningful question that humans can ask—and, no, fans of scientism ("scientismists?"), if a question can't be posed in a scientific way, sticking to the realm of the observable, testable, and falsifiable, then that doesn't automatically make it meaningless. Not every question that matters to us sits within the realm of empiricism and rationality that make science so fun. I half joked earlier about the nature of love and the nature of time as two quick examples, but take another beat to consider it. Can the methods of science ever hope to answer every question we might have about love? Do linguists, artists, anthropologists, philosophers, and religions have nothing satisfying to add to the conversation?

As for the nature of time? Well, if you want to wait for science to figure that one out, well then you can just insert your own joke about time here.

Look, I don't want to give the wrong impression here. I'm not saying that religion or any professed articles of faith should automatically take precedence over statements made through the exploration of evidence and rationality. I personally believe that science produces some amazing, beautiful, and *correct* stories of the universe—but what you make of the universe is your own business, and I believe that nobody can tell you otherwise. And to pretend that religion doesn't offer anything of value to the human race is to . . . I hate to say it . . . ignore the empirical reality of the world we live in.

The same goes for philosophy. Or art. Or history. We live in a complex, deeply interconnected world. It's beautiful, isn't it? Should we only look at this world through a single lens? Our ancestors had it right: the liberating arts, the knowledge needed to free the mind, includes all explorations of the universe.

I've observed that scientists are humble in a very strange way. I've lost count of the number of times that people, having learned I'm an astrophysicist, as soon as I say something wrong or make a mistake, exclaim with delight that *a scientist got it wrong.* But scientists in reality spend most of their working lives in the dark (sometimes literally): we're wrong more often than we're right, and we don't ever expect to be right unless and until the evidence shows it. Practicing science is a living meditation on the power of ignorance. Scientists should be the first ones to recognize that the world isn't simple and that answers don't come easily.

Armed with this deeper appreciation for the richness of the world of the not-science, we can start to navigate a change. We can insist on liberal arts educations for all students, especially science majors. We should include a correct discussion of the history of science and ground science education in the philosophy that grounds science. We can resist the segmenting of academic majors and the view that a chosen major is just a glorified job-training program.

We can encourage dialogue and collaboration among departments, recognizing the true power of the university: not in pumping out graduates as efficiently as possible, but in representing the best of the human intellect. What would a collaboration between an anthropologist and an architect look like? What fruits would it yield? I don't know—so let's find out.

Lastly, we need to get the public on board. For some reason or another, the majority of people believe in conflict between science and religion and have a distorted view of the institution of science. We can host public debates and dialogues, bridging the gap between campus halls and organizations intertwined in communities. We need to share; share perspectives, share methodologies, share viewpoints. Not with a goal of arguing or winning or convincing, but with the goal of building.

Scientists and science communicators need to highlight—*in a good way*—the shortcomings of the scientific approach. If more people understood what science is and what it can do—*really do*—then they can be free to trust scientists more and engage with them more. It's only through honesty that we can achieve progress.

6

The Lack of Diversity

We've got a problem.

UNFORTUNATELY, the most notable scientists who aren't white men are notable for becoming the *first* nonwhite and/or female scientists in their fields. For example, take Edward Bouchet, the first Black American to get a PhD in anything (which was physics, of course) from Yale in 1876. He went on to a difficult and tumultuous career in teaching because, despite having a PhD in physics from one of the most prestigious institutions in the nation, nobody would hire him because he was Black.[1]

It wasn't until the mid-twentieth century that we start to see Black scientists gainfully employed *as actual scientists*, like Harvey Banks, who got a PhD in astronomy and actually got to be an astronomer. Similar stories surround women in the sciences, like famed astronomers Henrietta Swan Leavitt and Annie Jump Cannon, whose work revolutionized our understanding of the cosmos but who were initially only hired as human "computers."

Since then there have been many successful minority and women scientists, many of whom have made important, significant, and lasting contributions to the field. But the problem isn't that minority and women scientists don't exist; it's the fact that there are way too few of them given their percentage of the American population.

In 2021, a total of 40,859 PhDs were earned in all science and engineering disciplines in the United States (and yes, we're very good at keeping track of these kinds of numbers). Of those, a mere 2,135 went to Hispanic or Latino scholars, 1,392 went to Black students, and 64 went to American Indian or Alaskan Native scientists. Of all the doctorates awarded that year, only 43 percent went to women.[2]

Some fields are far worse than others. For example, in 2011, the year I earned my PhD in physics, I was one of 826 students in the United States doing the exact same thing. But in that same year, only 17 of those students were Black.[3] In 2003, of all the PhD physicists working in American national labs, a grand total of . . . exhale . . . 11 were Black.[4]

The data for 2021 paint a stark and unequal picture across the other sciences. For example, Latinos represented nearly 10 percent of all PhDs in psychology but less than 2 percent in computer science. For Blacks, they were awarded over 11 percent of all doctorates in health science but less than 2 percent in computer science, engineering, Earth sciences, mathematics, and physical sciences. Native students made up less than 1 percent of all PhD awards in *all* fields, and in computer science and mathematics there were *none*.[5]

Every year there are a dozen subfields of math and science where *not a single Black, Latino, or Native student earned a PhD*.[6]

And while women made up over 70 percent of all psychology doctoral awards in 2021, they held less than a third of all diplomas in the physical sciences, mathematics, engineering, and computer science.

Even though there have been some gains across all these fields over the past twenty years, they haven't been remarkable. Black doctoral recipients increased from 2.6 to 3.5 percent of the total from 2001 to 2021. For Latino students in that same time frame, their proportion went from 2.6 to 5.2 percent. Native students suffered the worst. Despite the total number of doctoral students nearly doubling in the twenty years, the number of awards going to that group actually dropped, going from 0.2 to 0.1 percent of the total. Women also saw only modest gains, going from 38 percent of all science and engineering PhD awards in 2001 to 43 percent in 2021.[7]

Again, those numbers hold as averages across the sciences, with some fields doing (marginally) better than others. Just to highlight one field, over the entire forty-year period from 1972 to 2012 (these numbers are coming from all sorts of various studies, hence the apparent randomness of the years), there were a total of 28,859 physics PhDs. Of those, again in the entire forty years, 354 went to Black men and 66 went to Black women. Between 1977 and 1982, there were a total of 51 PhDs awarded to Black physicists. Between 2001 and 2006, there were 64. Think of all the revolutions in thought and technology that occurred between 1977 and 2006. Think of all that remained stagnant.

When it comes to bachelor's degrees, the same complicated yet dire picture emerges. Across all fields of science, there are indeed more and more young women and minority scientists entering undergraduate programs every year (in some cases increasing by about 50 percent in the past decade)[8]; there are also more and more young Everybody scientists entering those same programs every year, so the percentage of women and minority students doesn't always increase. For example, in physics the raw numbers are going up: from 158 bachelor's degrees awarded to Black students in 1997 to 263 in 2017, but the percentage is dropping, from over 4 percent of all physics degrees to about 3 percent. More students are attending college than ever before, but Blacks, as a group, are choosing not to enter science.[9]

From 1998 to 2018, women made modest gains in the social sciences, where they held over half of all new degrees, and in engineering, where they went from 18 to over 22 percent. But in other fields they slipped, dropping from 26 percent of all computer science grads to less than 20 percent and from nearly 47 percent of math degrees to just over 42 percent.[10]

Hispanic students made noticeable gains in the past decade, nearly doubling their representation among new degrees in every field of science. But Black students have fared far worse, with almost zero percentage growth from 2008 to 2018, and in some cases, as we saw, losses. We can also see the same divisions with PhD output: in 2018 Black students made up around 12 percent of psychology graduates but less than 5 percent in engineering and mathematics.[11]

Yes, Blacks, Latinos, and Natives make up minority percentages of the overall US population. But we should expect, at least naively and absent any other factors, that the overall percentage of these populations earning bachelor's degrees and doctorates (not to mention moving up the academic food chain afterward) should reflect roughly their overall representation in the American population. If that were the case, we would see *all* fields of science having roughly 50 percent women, 19 percent Latino, 14 percent Black, and 1.5 percent Native.[12]

Very roughly speaking, we're missing about half—if not more—of the expected population demographics in the sciences.

Like I said: we've got a problem.

Here's the thing. I'm a white dude from a middle-class background, raised in The Middle of Nowhere, Ohio. I don't believe that I received any special treatment or consideration. I had to take out student loans to pay for my bachelor's degree (which you purchasing this book is helping to pay off, thank you very much). I didn't get into my first choice for graduate school.

Did I have the hardest possible path to a PhD? Certainly not—I've met many colleagues from many backgrounds whose tales of how they got to where they are keep me up at night. But I can say without any sense of guilt that I believe I had to work my butt off to push through my career in physics. Nobody looked at me and said, "Hey, you're white, here's a pile of cash and some good connections."

On the other hand, and this is important: *nobody questioned why I was there.* I could walk into any physics department in the country (and heck, probably the world) and say, "Hi, my name is Paul. I'm interested in science," and nobody would bat an eye. Nobody would say, "Wow, I don't meet a lot of people like you!" Nobody would say, "Where are you from?" Nobody would say, "Are you in the right department?"

If I walked into any physics department in the country and looked around (which I pretty much have by this point), I would see a lot of people who look like me: a white dude from (insert random small town here). The folks in a typical science department not only look like me but they talk like me, act like me, see the same movies as me, listen to the same music as me, have family structures that look like mine, come from interchangeable neighborhoods, and most importantly for this discussion: they think like me (well, a much more educated version of me).

Sure, there were two Black professors at my graduate university (out of a hundred at one of the largest physics departments in the country), but heck, my graduate advisor *was literally named Paul.*

I'm not projecting any guilt from this statement; again, I still worked hard. But there was a certain assumption about me, that my interest in science and pursing a PhD could be taken at face value.

I'm telling you this because we build up biases without realizing it. We slowly—ever so slowly, and without ever thinking it directly—begin to assume that the way we see the world is the correct one. I mean, how many dinnertime arguments have you had with family members that eventually devolve into completely opposing worldviews and assumptions about the way things work?

As online forums (and later social media) would show, we tend to naturally group ourselves with similar-thinking people. After all, it's way more fun to find a support network than get in constant go-nowhere arguments.

Over the past hundred years, the academic expression of science has ossified—a giant social media hive mind dedicated to the pursuit of understanding nature.

At first, it was a matter of outright racism: Black people (and Latinos, and Native Americans, and women, and others) were considered intellectually inferior, because . . . well, *because*, and so they had no place in the hallowed halls of science. And as time went on, it became a self-fulfilling prophesy: only white men could advance in the field (let alone get admitted), so only white men made Discoveries of Great Importance, reinforcing the belief that only white men were the kind of folks able to do the job of science.

Nowadays, I feel confident in claiming without any hard evidence to back it up whatsoever except for the hope that humanity in general has a slightly better perspective on race than we did a century ago, that there are few-to-zero outright racists in the halls of academia (misogynists, well, that's a different story, which I'll get to in a bit).

Faculty at universities are educated people—by definition. And they're educated in more than their specialty; they receive some sort of modern, liberal arts(ish) education, which makes them generally aware of the current status of the world and informed about modern sensibilities when it comes to perceiving that same world.

Specifically, a typical scientist at a typical institution will be generally (if only vaguely) aware of what other typical scientists at other typical institutions are thinking about their particular topics. Including social scientists, neuroscientists, geneticists, and biologists.

My point is that a member of a scientific community is likely to be aware of the fact that "race" is largely a social construct.[13] That when you divide humanity along a certain set of arbitrary lines, you end up with arbitrary differences with no real significance or meaning except for the ones that we create in our imaginations.

Yes, there are ethnic groups and genetic lineages. And yes, a hundred years ago we thought that "race" meant something really important and really significant, so much so that we figured that skin color was enough to entirely classify a population of humans into anything remotely resembling a coherent system.

We also thought that the Sun burned coal to stay warm.

This kind of thinking was bolstered by IQ tests that showed "scientifically" that white people were superior. I won't get into the whole IQ test *thing* here; suffice it to say that for a long time we assumed that how you scored on this particular test was a measure of how smart of a person you were and how likely it would be that you succeeded in your life.

In the intervening decades between now and then, we learned a few things. We learned that IQ tests tell us more about testing and education methods than they do about human capacity for intelligence and creativity. We learned that if you don't give a population of people a decent education, they tend to underperform in IQ tests.[14] Huh, weird.

We also learned that racial dividing lines (assuming you can even get more than one person to agree on the same definitions) don't really mean much. You can divide people based on skin color or eye color or lactose tolerance or hirsuteness or eyelash length and you end up with . . . people divided along those lines and those lines alone.

It's true that a white kid with ancestral heritage from Poland is genetically different from a Black one with a family history from Ghana. It's also true that a white kid with ancestral heritage from Poland is also genetically different, *with similar if not greater magnitude*, from another white kid with ancestral heritage from Poland.

The genetic differences between ethnic groups (let alone "race") are no bigger than the genetic differences between any two random people. If you divide people based on a few differences, it doesn't tell you anything deeper about how different those people are.

Short version: Black and Latino and Native kids are just as smart as white kids, and girls are just as smart as boys. Period.[15] They are just as capable, just as intelligent, just as promising as their white male counterparts. They have all the same raw capabilities—grit, perseverance, curiosity, eagerness—that it takes to make a great (or even mediocre) scientist.

So where are the Black scientists? The Latino scientists? The Native scientists? The women scientists?

The issue is a subconscious, creeping institutionalism of discrimination, also known as implicit bias. The generic ivory tower intellectual is generally aware of and agrees with the above discussion because the above discussion is based on the evidence and scientific analysis, and scientists are mildly prideful about the superiority of their way of understanding the universe.

That's why I feel confident in claiming that there is little overt racism happening in modern science. But there's a difference between racism and bias.

It's certainly possible for a professor to give—again, likely without realizing it—preferential treatment to students who look and act and think like themselves, and to not give that same treatment to students who look and act and think differently.

There's a very sneaky mental sleight of hand going on here. Let's say you're a successful scientist, tenured professor and all (and maybe a few awards to your name). A lifetime of service to your field and your department. You know your life's story. You know the obstacles you faced. You know the challenges you overcame. You know how you, sometimes through sheer force of will, *did it*.

And then you meet a young student, asking for a position in your research group. If that student resembles you (or at least, the fantasy-based memory of what you thought you were like at that age), then you're probably more likely to think that this student has the makings of the Next Great American Scientist. Hired, on the spot.

But if they are different? Have a different background, or experiences, or attitudes? Well, you're not quite sure if they have what it takes because you have nothing in your personal inventory of evidence to compare it to. Sure, the kid showed up at your office door, which takes a certain amount of guts, but scientific research is (1) hard and (2) uncompromising. To continue to be a successful researcher, you have to hire good (if not great) students and research assistants. A bad junior member of your team can weigh you down like an anchor, delaying—and perhaps even damaging—your pursuit of scientific fame and glory.

A student is, essentially, a bet. And there are so many students knocking on your door that you generally have the luxury of picking among many candidates. With research money so scarce, every decision counts. Every dollar matters. When you're about to invest in a promising up-and-coming student—and it really is both a financial and mental investment for the first couple years—you need to make a safe bet.

So do you go with the student who looks like a young, eager version or yourself? Or do you take a risk? Generally, as we've seen in the dreadful statistics above, you make a safe bet.

And this begets a monoculture in science: senior scientists hiring junior scientists who look like the senior scientists, who then go on to have successful careers and fill the faculty offices with a new generation of senior scientists, and the cycle continues.

And I'm not just talking about white scientists here. People of Asian descent are overrepresented in the sciences with respect to their share of the American population.[16] Why? Because (1) there are a lot of Asians in the college system, and (2) by now, there are a lot of Asians in science departments across the country. There's enough "evidence"—in the form of verifiable track records that Asian students make for safe bets—that hiring them into junior positions isn't a risk.

It's a nasty expression of the risk aversion that we've explored elsewhere in these pages, a risk aversion that has turned into a desire to keep things the way they are because it's too expensive and risky to attempt change. It's an avoidance that doesn't target certain programs or paths of scholarship but rather entire groups of worthy human beings.

In essence, anything "nontraditional" hurts your chances of becoming a long-term successful scientist, where "nontraditional" can mean the wrong skin color, the wrong gender, the desire to raise a family, or even an interest outside the mainstream. In the race to the top of the science food chain, there's a certain default assumption of what a "brilliant" scientist looks and acts like—namely, a white dude.

Science started as a white man's game centuries ago, partly because of the historic origins of science (i.e., Europe) and partly because of overt racism. And over time, without anyone trying, that pattern continued, right up to the modern era. This monolithic culture perpetuates itself, as these kinds of assumptions infect all levels of the scientific education and training program—the whole path from high school student to Nobel Prize winner. In other words, the entire pipeline, from high school to undergraduate through a PhD and onward to an illustrious career, is completely broken for minority and women scientists.

The problem starts at the beginning of the funnel. Young (as in, very young) minority kids and girls have very few role models to look up and aspire to. And if you think that's No Big Deal, just ask yourself if, as a kid, you looked up to any role models and sought to emulate their careers. Did they . . . look like you, just a little bit?

I can relate to a small sliver of this sentiment: when I was just a little Paul, I absolutely loved all things science. I would gobble down books as fast as my mind could absorb them. But because I didn't know any scientists, was living in a random midwestern town, and nobody bothered to actually inform me of the possibility, I just assumed that science was done by "somebody else."

So I went on to start a degree in computer science (because I was also, and continue to be, a huge computer nerd). It wasn't until my third year, when I took an astronomy elective class, that my professor pointed out that this thing called "physics" is actually a viable career and that I might enjoy it.

If I had a hard time finding a path to science despite my natural inclination toward it, what are the chances that some random Black, Latino, or Native kid is going to sign up for Astro101 at the nearest university? Statistics bear it out: not very likely at all.[17]

There's a scarcity of disadvantaged kids who even contemplate the possibility of a career in science. And once that's ingrained it's pretty hard to shake off. As fewer and fewer Black, Latino, Native, and women students make it into universities, get degrees, and dedicate their lives to science, there aren't any professors *who look like that* sitting around, ready and eager to mentor and hire more young minority and women students. With few minority and women scientists, they aren't appearing in public outreach events or TV documentaries, giving young minority kids and girls someone to look at in the scientific field and say, "Hey, wow, that person looks just like me and is a scientist. . . . I wonder if . . . I . . . could be a scientist!"

From there, high school teachers inadvertently discourage underrepresented kids from going into science, assuming that the mathematical prowess of those kids isn't up to snuff, suffusing every single interaction at those tender ages with an air of *you're not good enough.*[18]

Research has found that a perceived gender bias is able to predict an underrepresentation of women getting bachelor's degrees in certain fields—the more that women feel that a field isn't "for them," the less they sign up for class.[19] And when they do get to class, a "benign" (I use that word loosely) form of bias where male teachers act more paternalistically to female students hurts young women's GPA, especially if they don't already strongly identify as a scientist.[20]

From there, fewer women enter fields whose participants are perceived as "brilliant"—a descriptor typically reserved for the masculine.[21] And that stigma follows them throughout their careers: for some reason, women constantly have to choose between the identities of "woman" and "scientist," whereas men are allowed to be both.[22]

The concept that women and minorities aren't cut out for science is one of the most absurd arguments I've ever encountered. These people have literally ruled nations, and yet a double-blind study is somehow beyond their reach? And so women and minorities leave science in droves not because they're not cut out for it, but because of the hostile attitudes that are consistently allowed to exist.

And even when those kids do make it into college—despite all the pressures against them—they often don't end up at the "right" school. A scientific career is based on networking: knowing the right people at the right time. If you don't go to a top-flight university, your chances of making it into a top-flight graduate school are slim. And if you don't get into a top-flight graduate school, the top-flight researchers at those institutions don't know who you are. And if they don't know who you are, they're not going to hire you. And thus the system finds another way to perpetuate itself.[23]

And if an underrepresented kid does manage to get into a good college or a doctoral program, what will he or she see when looking around the department? Black faculty? Latino grad students? Female undergrads? Anybody—*anybody*—to serve as a mentor or advisor?[24]

In their departments, women and minorities systemically get fewer mentoring opportunites.[25] As they continue their careers, they then face an unholy gauntlet of higher standards, where they need to be *more ideal* than the ideal (white male) candidate, making it even harder to pass the bar to the next stage of their careers.[26] And while some fields are reaching gender parity, overall women do not advance as far or as frequently as men, are not cited as much, and do not get invited to speak as much.[27] In other words, they don't get to do all the fun things that successful scientists get to do.

The whole societal attitude toward work/life balance generally hurts women (who are expected to raise children) more than men (who are expected to keep working) in the professional world, and naturally this insidiously prevents women from advancing as far or as fast as their male colleagues in the sciences.[28]

At the other end of the career spectrum, you don't have to look hard to find examples of women and minorities snubbed for high honors and praise, especially when it comes to the Nobel. Only a measly 6 percent of those prestigious prizes have gone to women, despite their awesome achievements.[29] And that 6 percent counts the Peace and Literature Prizes, which are very nice awards but not the subject of this book. In Physics, Medicine, and Chemistry, of the 624 awards given in those categories since Alfred Nobel made the whole thing happen, a grand total of . . . 23 . . . have gone to women. Zero have gone to Black scientists.

Indeed, even when women and minorities do achieve as much (or more!) than their peers, they report higher levels of self-doubt, something called in academic circles "imposter symdrome."[30] Despite all evidence to the contrary, many women and minorities can't shake the nagging feeling that they're simply not as good men (I wonder if it has anything to do with the hostile cultures and double standards). In fact, we don't even have a solid idea of how many women and minorities are "leaders" in STEM, and every place we try to look for some vague handle on the problem, we find those groups grossly underrepresented.[31]

This results in Black, Latino, Native, and women students facing a disproportionate uphill battle in their quest for a science career.[32] They drop out of STEM majors at twice the rate of their white counterparts. They tend to borrow money at higher rates, usually to pay for more expensive educations at for-profit institutions. Programs developed by government agencies trying to get minority students into science careers go largely untapped.[33] Even established minority and women scientists have a harder time than their white male competitors in winning grants.[34] It just doesn't stop.

Sometimes it's not just implicit bias and quiet discrimination. Sometimes it's outright harassment. The sciences are a white-male-dominated enterprise and have been for a long time. It's also a culture with extremely high imbalances of power, where senior faculty have the ability to make career-defining (and career-killing) decisions regarding their juniors. If you don't just go along with whatever your advisor thinks is cool, then you're in for a rough ride. And "whatever your advisor thinks is cool" can run the gamut from interesting research areas to sexual harassment.

In short, women and minorities in science find a culture where harassment, bullying, victimization, and abuse are simply *permitted*.[35] Outnumbered in labs and offices across the world, they find themselves the target of negative stereotypes that they can't fight against.[36] Yes, there are bullies everywhere, but the way science is structured allows jerks to flourish without restraint and without consequences. I've been extremely lucky—none of my personal advisors or mentors has been a horrible person, at least to me—but I've certainly encountered more than my fair share of random jerks in my work, and all too often women find themselves easy prey for their abuse. They're just such easy targets, because every time they go to defend themselves, the (jerk) men in science claim that their attitudes and behaviors are *just the way it is, sweetheart*.[37]

It's the normalization of the behavior that hurts the most. The claim is that science is tough and exacting, and so senior researchers need to be hard on their juniors so that they can succeed—and it *just so happens* that jerks get rewarded and nonjerks (i.e., normal people) suffer, along with any minority group trying to play the science game. Academics have long had cultures of open secrets (as in, "Oh yeah, that guy is totally a sexual harasser"), superstars (as in, "Oh yeah, that guy is also beyond famous and gets tons of grants so we can't touch him"), and a bizarre blending of professional and personal lives (as in, "If he asks you out to dinner, just say yes because it's technically a work thing").[38]

And so we end up with people like UC Berkeley's Geoff Marcy, an expert at finding planets orbiting other stars, a frequent short-lister for Nobel Prizes, and harasser of countless women throughout his career—*all while everybody knew what was going on*.[39]

The severe and often debilitating power imbalance in academia crystallizes with that all-important key ingredient for scientific career advancement: the letter of recommendation. There's sadly very little research on the topic because there are so few minorities to build statistically significant numbers from, but we do have some insights. One study compared letters of recommendation for white and minority students with similar GPAs. The study found that, overall, white students got longer, more detailed letters, including detailed accounts of the student and descriptions of their productivity. On the other hand, minority students received shorter letters that emphasized the "insight" of the student (and other such vague terms) and included more hesitant, tentative statements about their actual abilities.[40]

The end result: despite quantitative similarities (like their GPA), minority and women students were spoken about in less-than-glowing terms, resulting in fewer job placements for those students.

It's subtle, of course, but when the difference between an amazing and a mediocre letter of recommendation can mean the difference between a lifelong career in science and a lifelong career in, well, anything but science, those tiny little variations in the wording (measurably) add up.[41]

This doesn't even bring into consideration the interrelated problems of race relations and poverty in the United States. So now I'll bring that up. Minority communities are disproportionately poor communities, and poor communities worldwide have a hard time advancing their young men and women into scientific fields. A poor white kid from a defunct Appalachian mining town will be unlikely to find a pathway to a Nobel Prize, but there are plenty of white kids *not* from defunct Appalachian mining towns to make up for that.

When it comes to a Black kid from the wrong side of Brooklyn, they're not likely gunning for a free trip to Stockholm either. And that's the end of the line—there are (essentially) no other Black kids from wealthier backgrounds to take a stab at it.

So here we see in science the end result of decades of systemic, and in some cases institutionalized, racism and discrimination in America. Careers in science are simply not an option for poor kids (which is a serious travesty, even disregarding the race issue), and this affects minority communities more than white ones.[42]

And heck, even if a young, disadvantaged genius gets a foot in the academic door and is able to wind their way through the beginning stages of a scientific career, they have two choices: go Full Science, with its long hours and low pay, or take that shiny new PhD somewhere else, where it can actually earn some serious coin.[43]

Being a scientist is a luxury. Entire civilizations need to attain a certain amount of wealth before they can go from "okay, fine, we'll pay for one astronomer to make sure we start the spring harvest on time" to "okay, fine, we'll build a particle collider so you can understand the fundamental interactions of the physical universe." And what applies at the macro goes for the micro: what we currently see as a lack of diversity in science is really the aggregate of millions upon millions of individual societal choices.[44]

Not only does a disadvantaged kid in a poorer neighborhood have very little chance of even realizing they can have a career in science, if they go down that road they're going to have a tough time of it, and even then the potential rewards don't really outweigh the costs. If you want to argue that Black, Latino, or Native American culture disfavors or disapproves of science, then you have to explain *why* that's the case, and once you follow enough *whys* you tend to end up right back where you started: at some form of discrimination.

At all levels, the STEM training and job career path fails, with young minority and women scientists exiting their careers at a far faster rate than their white colleagues.[45]

It's a trap, a trap fueled by a refusal to change the status quo.

And here's the danger of all this monolithic ossification in science: *monolithic ossification*. Science is, by its very nature, a deep look into the very heart of nature. It's not easy. It takes curiosity, creativity, guts, passion, and perseverance. Nature hardly ever reveals its secrets—and if she does, it's usually only in the faintest of whispers.

I mean, come on, look at something like the history of understanding the atom. It took over a hundred years just to convince our scientific selves that atoms even exist. But after countless experiments, observations, and arguments, we were able to convince ourselves of the existence of something that we literally can't even see. On the other side of the cosmic coin, nobody was around in the earliest days of the Big Bang, and yet we're able to calculate with confidence the amount of hydrogen and helium in the universe.

You think that's easy?

This is where diversity matters. Not necessarily a diversity of skin colors or sexes or orientations, but what those physical differences represent in the thinking patterns and thought processes of the wonderful brains inside those bodies.

Lots of people can be smart, but not all smart people can be scientists. It takes a certain *kind* of smart to be a successful scientist; the kind of smart that marries intelligence with a deep curiosity and personal approaches to problem-solving. There's no real difference—statistically speaking—between the raw intelligence of a random white man and anybody else, and so on the surface it may not appear as if science is hurting because of a lack of diversity. But a person's ability to make intuitive leaps of understanding or see subtle connections where none are obvious is shaped by more than raw intelligence. It's shaped by their personal story, their history, the stories they heard growing up around the dinner table, their culture, their values, their faith, their community of friends.

Understanding the world around us in a scientific way is really hard, and the dangers of a scientific monoculture can make it even harder. If everyone thinks the same way and tackles problems in the same manner, then it becomes ever more difficult to actually think of new things.

You may wonder why many fields of science nowadays are only able to make incremental progress. Part of that reason is the vast complexity of the universe around us, and the reality that all the problems that could be solved by a single person sitting around in a dusky office thinking about it have already been solved. But perhaps part of the problem is because the scientific culture has created a distinct lack of viewpoints: all scientists thinking the same way about questions that they all agree are important.

The pervasive systematic biases in science harm not just nonwhite, nondudes wanting to enter science but also all the "traditional" scientists . . . and even science itself. The more the culture of science and academia resists change, acceptance, and openness, the more society will put up mental (and financial) barriers to the successful integration of science into the everyday world.

If science as an institution is going to survive—and thrive—well through the twenty-first century and beyond, it's going to *need* (as in, actually fundamentally require) to cultivate a diversity of viewpoints and brilliant minds. And the fastest way to have a diversity of minds is to have a diversity of bodies, with different backgrounds and family histories and communities.

And what's more, and this is a thought that particularly troubles me when I can't sleep at night: how many Einsteins have we left behind? Old Albert's mind was truly unique—he was capable of making astounding leaps of insight that literally *nobody* else could think at the time (or even since). Sure, some of the things he dreamt up—like special relativity—had other people thinking along similar lines, but other things—like his magnum opus, general relativity—might not even exist today, a century later, without his genius.

Einstein was also very well positioned: he had a good family, he went to a good school, he lived in a good part of the world (well, at least for a while), and he lived in a good time when that part of the world was heavily investing in scientific research.

How many kids that have Einstein-level abilities to solve some of the most pressing questions in modern science end up in nonscientific fields (hopefully at least serving the advancement of the human species in some capacity) or in some dead-end rut? How many kids have the never-activated potential to push a field of science into an entirely new regime? How many kids never see a future for themselves in science, and how much are we suffering because we ignore them and push them away?

We owe it to our children to create a more equitable system. Do we really want to live in a world where nearly half of the human population doesn't get to participate in science? Where their own creativity is quashed? Where their own dreams don't count? Do we really want to turn away from all those geniuses, all those dedicated workers, all those sharp minds? Do we really think we're fooling anybody—let alone ourselves—that we can solve all these damned hard questions of the universe without as many minds and hands and hearts as possible tackling them together?

That's the bad news. The good news is that we have a chance to fix it. But, as you might've guessed, it's going to be hard. You don't correct over a century of racism, discrimination, and bias with a few bullet points. It's going to take work and, most importantly, risk, especially at the institutional level. We can't put the onus of solving these problems on disadvantaged kids trying to get ahead in life and fulfilling their dreams of becoming a scientist—they already have their work cut out for them.

The first step is to celebrate minority and women scientists, giving them access to platforms for science outreach—the more they appear and speak passionately about their passions to rooms full of kids (and sometimes more importantly parents, who have a not-so-subtle influence on the future trajectories of their children), the more those kids will identify role models and potentially realize that they could be scientists too.

There's very little connection between the world of professional academia (professors and postdocs in cahoots with their grad and undergrad students) and the places those younger students come from; namely, high schools. Occasionally—and I mean "so occasionally it basically never happens"—there will be mentoring relationships built, but those are only available for the brightest/most gifted/most well-connected students. Throughout my career, I continually get requests for shadowing experiences, where a high school student wants to follow me around for a day to "see what it's really like to be an astrophysicist." I decline almost every single request, because the reality of being an astrophysicist involves typing emails, sitting in telecons, attending meetings, and staring at a computer screen that refuses to give the correct answer.

In other words, following me around would be perhaps the greatest turnoff to science for a high schooler possible.

But there is more to the science world than theoretical astrophysics (I think). There are real-life labs filled with people wearing real-life lab coats doing real-life experiments. There are giant supercolliders with all sorts of gizmos that could make even the most ardent engineer sweat. The machine of modern science is breathtaking in its scope—and it's almost never shown to disadvantaged kids.

It sounds trite—let's give girls and minority youths some more tours of our labs and facilities—but the smallest spark can ignite the biggest of fires, so it's worth a shot. Mentorship is one of the most powerful accelerators of kids, capable of literally transforming lives.[46]

Next comes outreach, not only in the form I just talked about but professional outreach. As in communicating to underrepresented kids about the *value* of a degree (and, optionally, a career) in science. Physics and astronomy PhDs have basically zero unemployment, which is an attractive thing to hear if you're trying to climb the socioeconomic ladder. Simple facts, laid out clearly and concisely, broaden the horizons of the typical undergraduate science applicant pool.[47]

We need to understand their identities and priorities and stop thrusting the standard This Is a Scientist character onto them and hope it all works out for the best. In other words, we need to meet disadvantaged kids where they are—who they are—and allow *them* to grow into their own.[48] Sometimes it's even as simple as ensuring that Black, Latino, and Native students are *welcomed* as valued colleagues, which for some reason isn't simply happening as much as it should.[49]

Since most disadvantaged kids who do move into education beyond the high school level do not appear in colleges, we need to stop assuming that those kids who have a desirable scientific skill set will simply appear on the Quad. They tend to go to community colleges—because they believe, by and large correctly, that that is the fastest route to gainful employment. So the very next Einstein could be learning how to fix car engines, which isn't a bad job at all, but there are plenty of car mechanics out there in the world and not so many Einsteins.[50]

Researchers, faculty, labs, and universities need to build relationships with community colleges, creating bridge programs that allow interested students to audit classes, meet with scientists, and even participate in lab life. In surveys, minority scientists have frequently cited access to scientists at early career stages as helping to propel them forwards.[51] In short: we need a better supply system.[52]

Within the halls of universities, the culture of academia must shift away from the attitude that if you're not devoting every waking second (and most of your sleeping ones) to the pursuit of science, then you're not worthy. We've already seen the damaging effects of this on young scientists of all kinds, and it disproportionately hurts women and minorities. Science as a job needs—*needs*—to become family friendly, allowing healthy work/life balance and pursuits outside academic hallways.[53]

Within individual institutions, research has shown that more structured organization, identification, and dissemination of common goals and a rich culture of mentorship help keep a lid on bullying and abusive behaviors, all of which help retain women and minorities who choose to pursue careers in science.[54]

With the most egregious forms of abuse and harassment stamped out, with violating faculty actually facing serious consequences like loss of tenure (which eventually happened to Geoff Marcy, by the way), there's still a long way to go. Departments need to actively seek out, recruit, and advance minorities and women—they can't just cruise along hoping it will all work out in the end.[55] Strong peer mentorship initiatives helps retain talent.[56] Scientists of all kinds and all career levels must project a sense of welcome and support, and must actively fight against "this is a scientist" stereotypes to allow young women to identify both as women and as scientists, something that men can do with ease.[57] For example, programs that have confronted social stigmas head-on have found a lot of success in helping to transform the lives of young women and minority scientists.[58]

There's also the minor fix of *not letting abusive jerks run around with impunity*, which shouldn't even have to be stated, but here we are.

Lastly, scientists need to take *risks*. Every potential research assistant is a gamble. Either they end up becoming productive and self-sustaining, capable of churning out valuable research in your name, or they don't. The risk aversion prevalent throughout science—due to the relative lack of funds and the relative abundance of warm bodies—makes the science industry a "buyer's market." Group leaders get to take their choice from many promising, eager, enthusiastic candidates.

And maybe they shouldn't. Maybe top-flight researchers should deliberately hire and train candidates who they believe aren't the best of the best. Or more accurately, students who have all the right *qualities* of being a capable scientist but not necessarily the right *training*. All too often, minority and women graduate students and postdocs are seen more as a burden than a blessing, which harms not only their chances of persisting in their careers but their emotional health as well—this is all *despite the fact that they're performing just as well as anybody else.*[59]

And as for the students who have both the right qualities and the right training? Something tells me they'll be just fine in life, because smart, capable, hardworking people are always in demand. The fact of the matter is that with so few positions in science and so many students, someone's going to get stuck on the no-science-for-you track, despite their best intentions. Maybe it's time to stop being lazy and actively go out looking for promising minority and women candidates who might not give you the highest research ROI in the short term (not because of their lack of intelligence or creativity or grit, but because of the shortcomings of their education or background) but who enliven and enrich the scientific enterprise for all. And maybe we should hold ourselves accountable to these standards.[60] All too often, faculty make assumptions about the capabilities of their younger colleagues, without actually bothering to allow those colleagues to grow and learn, so any bias that the professors bring into the hallways of universities simply sticks.[61]

Ultimately, everything I've listed here as potential solutions barely even scratches the surface because, as we've seen, the lack of minorities and women involvement in the sciences is part of a much larger narrative. Teachers, mentors, hiring committees, grant managers, and active researchers all play their role in preventing those kids from succeeding, but they are as immersed in the wider culture as we all are. That said, baby steps are still steps. Small improvements in teaching, openness, mentoring, opportunities, and proactivity can go a long way.

Perhaps—and I'm just tossing this out there—the reason minority and women students choose not to go into the sciences is because they perceive the sciences to be hostile to them. And they're right. So maybe if we take active steps to make it even *slightly* less hostile, we may see more young diverse students earning science undergraduate degrees, applying for grad school, and becoming scientists.[62] And who knows, we might just start finding all those Einsteins out there.

7
The Disregard for the Public

This is going to get ugly, because for this one in particular . . . *it's personal.*

YOU'VE PROBABLY NOTICED BY NOW that I'm *slightly* passionate about science outreach and communication. I love sharing all the cool and wondrous and amazing things we've learned about the universe through science, from the guts of atoms to the birth of the cosmos. That love of sharing science has guided me to where I am today: writing this book to you, talking about the places where science is broken.

Science communication? Outreach? The simple and humble act of *sharing* what we know with audiences willing and eager to listen? A talk, a post, a quote?

Most scientists don't even bother.

It disgusts me.

Most scientists, at all career levels, want nothing to do with speaking to the public in any form. They don't want to give an interview with a journalist. They don't want to say something on social media. They don't want to speak to a group of strangers at a bar.

They don't want to share. They want to keep all their knowledge, all their expertise, to themselves.

I've had my own fair share of detractors from within academia. Faculty members telling me that what I am doing is useless. That it's a waste of time. That it doesn't matter. That money, effort, and resources should be spent elsewhere. I had to fight to justify my existence within an academic sphere—and continue to do so today.

But the hardest part isn't the people who want me to stop; it's the vast, vast majority of my colleagues who just don't care. Communicating science to the public is simply not a priority for most scientists, as evidenced by the fact that *it barely ever happens.*

For years I've asked myself why. *Why?* Why aren't my colleagues, who are just as driven and passionate about science as I am, interested in letting anybody else in on the secret? Can't they see the spark in the kid's eye when they fall in love with science? Can't they understand the value of promoting science to the general public—the same general public that is paying their bills? Can't they enjoy the fact that people really do look up to and admire scientists and that public outreach enhances that?

Don't my colleagues see how utterly transformative science communication can be? Can't they see how communicating, *connecting* with the public, is fun and powerful?

Apparently not, as I learned the hard way.

Scientists are surprisingly social and communicative creatures. We write emails. We draft papers. We give conference talks. Science works through the constant sloshing of ideas. Scientists have all the skills available to communicate effectively and often to the public, and they're perfectly willing and able to communicate their knowledge to their peers. So lack of raw skill isn't it.

And yes, scientists are busy, with most of their time taken up by writing grants, directing junior researchers, teaching, and other various academic duties. But most human beings on the planet are also busy, and there's nothing especially unique about the kind of business that scientists exist in. "Busy" is just the twenty-first-century shorthand for "I don't want to." If a potential donor walked into the department waving around a giant sack of money, looking for interesting projects to invest in but wanting to learn a little bit more first, I'm pretty sure that scientists in halls around the world would suddenly find their schedules miraculously clear.

Besides, science outreach isn't exactly a demanding task. You talk to one reporter for a half hour. You go give a public talk and get some free food and booze. You let some kids ask you about black holes or dinosaurs (doesn't matter what your specialty is, you'll get those questions).

Science outreach and communication isn't hard, folks. And if you ever hear a scientist complaining about how difficult it is to translate their research into something the nonscientist can understand, you can remind that scientist that they are generally considered to be among the smartest human beings our species has ever produced and regularly confront some of the most difficult questions ever posed by humanity—*with success*—so maybe, just maybe, they should suck it up.

I mean, you should say it more politely than me, but you get the idea.

So what's stopping them?

Before I dig in too deeply into my dissection of modern science communication, we need to address that fact that scientists do, actually, truthfully, earnestly interact with the public.

Sometimes.

So let's take a brief survey of what science communication looks like in the twenty-first century.

Perhaps the single greatest instrument that scientists use to communicate with the public isn't through direct, personal, intimate experiences but through the intermediary known as the *press release*. The press release is a simple and straightforward affair, and it goes like this:

1. A university or lab scientist does something of note, or something that can be construed as a reasonable facsimile of noteworthy.
2. A university or lab press officer interviews said scientist, tries their best to understand the cutting-edge research, pulls some juicy quotes, and writes a press release.
3. The release is, well, released to swim in a vast, frothing sea of other competing press releases.
4. If the press officer wrote a good headline (usually involving firsts, bests, largests, smallests, uniquests, or any other superlative), then the press release gains traction.
5. Good journalists will reach out for additional interviews and a greater understanding.
6. Bad journalists and robots pretending to be journalists will just copy the press release wholesale.

7. The general public reads the headline of the press release. A small percentage then go on to read the first paragraph. An even smaller fraction that then go on to read the rest of the story.
8. Science is communicated.

The vast majority of the time that the public encounters the thoughts and minds of scientists is through this format: filtered, diluted, edited, and contextualized. And it's through this structure that a typical scientist will feel most comfortable and familiar. The press officer is an employee of their university or lab, not some foreign, exotic, and untrustworthy "journalist" that they don't know. The press officer is used to speaking to scientists and distilling what they say into something comprehensible. The university or lab does all that vague stuff that it takes to actually do the job of communicating science to the public so the individual scientist doesn't have to invest all that much time or interest.

It's a machine, and it works, at least in its own way. I don't have a citation for you that counts how many science-based press releases exude from universities and laboratories worldwide every year, but let's just call it *a digital truckload* and get on with it.

Speaking of journalists, that's the second most popular venue for connecting scientists to the public. Triggered into action by a press release, an interesting newsworthy event (like, say, a rare conjunction of the planets or a global pandemic), or a good old-fashioned hot tip, a journalist will call or email the main department phone number looking for an interview.

Assuming someone is actually behind the desk, usually what happens is that a frantic email will go out looking for volunteers. It doesn't matter who—they just need a heartbeat and a university affiliation. Interview booked, in comes the crew, the call, or the email exchange to conduct the interview. Over the course of the interview (and I've done a lot of these), the primary goal is always to get the interviewer to understand the topic at hand—science communication of the classic variety, with an audience of one (or two if the camera operator or editor is paying attention). Interview over, the journalist then finds the best quotes possible (sometimes only one—once, in reference to staring unprotected at a solar eclipse, a half-hour interview summarized with me only saying "your eyeballs will melt") and puts together a package consisting of their own thoughts, the quotes from the scientist, and some other related information.

When you turn on your favorite news station, listen to your favorite radio program, or open up your favorite newspaper to learn about science, that's how that particular sausage got made.

I also do not have a citation for the number of news articles, TV pieces, and radio segments created every year discussing science. I assume it's also on the order of one or two *truckloads.*

At a higher level of production value sits the science TV show. Not just a slapped-together interview, these things cost *real* money. They are almost always scripted by nonexperts (yes, even the ones that feature real scientists), with varying levels of quality and desire to actually communicate the facts of science. Some shows are very solid, distilling what we know in a clear way. Some shows are middling, mistaking fringe theories for valid ideas. And some are just bad. Deliberately so—more interested in being entertaining than having real content. But due to their expense, these kinds of documentaries and shows are rather few and far between.

Far below media lies everything else. Sometimes departments, colleges, or science museums will host a lecture. Sometimes scientists themselves will organize events. When they do, they're usually pretty creative since the people involved do not have the time or budget to effectively execute a traditional marketing campaign. I've seen "Astronomy on Tap" talks at bars, "Scientist on a Subway" outings, and even a version of prison ministry (but for science education).

In your city there's probably something science themed and educational happening this Friday. Look it up.

There are also so-called informal educational institutions—science centers, museums, even zoos and aquariums qualify. If you have kids and live in at least a midsize city, you've probably found yourself and your family at one of these places on a random Saturday afternoon, wondering why the cost of the concessions is so high. Outside of school itself (with its "formal" education), these are places where youngsters are most likely to interact with science.

And then there are the schools themselves. Full of precocious, bored kids. With the exception of high schoolers, every single classroom visit I've ever done has been amazing, probably because the kids are *thrilled* to have someone new talking to them. I personally usually don't even show up with a talk: I just let the kids hit me with questions. It always works, and it's always hilarious.

That leaves social media. Blogging and podcasting, tweeting and Facebooking, Instagramming and TikToking. Of all the forms, blogging is by far the most popular among scientists who dare venture into the digital realms. The reason is simple: a blog entry looks a lot like a lazy paper, and if you spend a solid fraction of your life writing meticulous papers, then a blog post is like a minivacation.

It's in the form of blogs that you'll usually find scientists (at least, the ones willing to do this) in their most unfiltered forms: their words, their thoughts, their ideas. Direct from their brain to yours, via the internet. More on that later.

As for the very social of the social medias, scientists are just like any other person: full of opinions and very willing to talk about them. Chances are that someone you follow is already a scientist, and they'll be talking on social media about all the things that they want to talk about: their job, politics, memes.

All this sounds like a lot of science communication, doesn't it? All sorts of scientists speaking through all sorts of different channels, each adding their voice in the grand symphony of educating the public about what we've learned.

It is a lot.

It's also pathetically little compared to what it could be.

It's also broken.

Let's start with the press release. Supposedly a humble instrument for communicating and disseminating science to the public, its true intent—and the very reason it exists—is advertisement for the university or lab sponsoring the research.

Seriously, this is the only reason that press releases exist. It's not out of some sense of social obligation, or duty to the public, or just warm fuzzy feelings. It's cold, hard cash. When a press release goes viral, copied and recopied to news outlets around the world, it's essentially free advertising and name recognition. More awareness in the public equals more prestige for the institution equals more opportunities for donors, grants, students, and other rivers of money to flow into their coffers. All it takes is the (sometimes part-time) salary of a handful of press officers and you're set.

It's that cold, naked calculation that drives the choice of what to release, and why the vast majority of science never makes it into the public sphere. The most powerful science is often the most boring: not the eureka of a new discovery, but useful technique. Not a surprise, but the slow grinding accumulation of knowledge. But the science that makes for a good press release (or can be twisted to become a good story) is like a gold mine. A good press release that hits worldwide distribution can easily be worth a million bucks or more in equivalent advertising, but in order to have that reach, the press release needs to be good.

But what "good" means for a press release is far different than what "good" means for science and what "good" means for science communication. And under that metric, most science stories are simply *bad*.[1]

"If it bleeds it leads," the saying goes, but unfortunately for the media nothing much bleeds in science. And so the best/worst science press releases are led by the best/worst science headlines, which usually involve some use of obviously over-the-top exaggeration, easily refuted statement, and bold exclamation. Sometimes it's even all three wrapped up in one.

Let's check out some especially juicy ones:

"Bacon, Processed Meats as Dangerous as Smoking"[2] (They're not.)

"Beards Are Covered in Poop"[3] (They're not.)

"Your Nail Polish Might Be Making You Gain Weight"[4] (It's not.)

"Climate Change Is Causing Lost Sleep"[5] (It's not.)

"Harvard's Top Astronomers Says an Alien Ship May Be among Us"[6] (It's not.)

"Cheese Isn't Bad for You"[7] (It is . . . sadly.)

Once you see these kinds of headlines, you can't unsee them. And if you do go on to read the article, just remember that the process of science is slow, meticulous, almost always mundane, usually wrong, and never knows the answers ahead of time.

Everything that I'm saying here about press releases goes for general news stories and TV show too, from the podcast with a dozen listeners to the show with an audience of millions.

Almost every journalist, host, editor, producer, key grip, or whatever position people can take in the manufacture of media is simply not a scientist. That's not a knock: in my own experiences working with media, from interviewing with press officers to hosting TV shows, I've had the pleasure or working with so many smart, quick, intelligent, well-read people.

They're not scientists, which isn't a bad thing in and of itself, but they're charged with the main task of science communication nonetheless. And that's not exactly the job they trained for or the one they signed up for. Journalists and TV show producers want to tell compelling stories first and communicate accurate science second. Or third. Which is *fine*, as long as there is a healthy, productive, collaborative relationship between them and the scientists. Yes, there are bad journalists and greedy producers pushing schlocky, incorrect, inaccurate, bad science stories out there, selling them to anyone who will buy them. There are also a lot of good-hearted ones, wanting to entertain audiences without making a sickening funhouse-mirror version of science. Those people are also busy, tired, and underpaid. Their hopes and ambitions often don't match the realities of strict deadlines. It's a miracle that they are able to write or produce anything resembling good science at all, and I applaud them for their efforts.

It's almost impossible for the general public to tell the difference between a legit science program or article and a garbage one. And why should we, as scientists, expect anyone to tell the difference? Both the good and bad science media have all the same jargon words, all the same slick graphics, all the same authoritative voices. From an outside perspective, good and bad science media look indistinguishable. We can write article after article, and book after book (like this one), arming nonscientists with tools to spot bad science TV shows, bad science writing, and bad science itself, so they can better filter it out and remain pure.

Do we seriously expect it to stick? For every tool we give the general public, the shadier side of science media simply extrudes another dozen shows or articles, each one a pulpy mash of legitimate science, poorly understood conceptions of the natural world, a crackpot view or two, and slick storytelling that cares less for accuracy and more for Excitement™. It's a hopeless competition—the race is over before the gun has even gone off. It's so easy and so efficient to ignore good science in favor of good storytelling, and the simple profit motive will always win.

It's so easy to blame journalists, producers, and press officers for these bad headlines and horrible science shows. They are, after all, the ones who are writing and creating them—even if they have good intentions.

But the science headlines and shows can't exist without the scientists themselves. The scientists are the ones who are providing the quotes. The scientists are the ones who are letting themselves be guided by an interviewer to drive up the hype. The scientists are the ones who are feeding the machine. For every single example of bad science journalism out there, from a schlocky TV show to a nonsense article, there is a scientist behind it.[8]

Sometimes it's unintentional. Producers or a journalist approach a scientist. The scientist does the interview, giving it their best effort. The producer or journalist then go on to twist their words, edit them, and spin what the scientist said in order to give the prewritten narrative the authority it needs to sustain it in the public's eye. A nasty surprise, to be sure, and it's caught many of my colleagues off guard. While the fault lies largely with the duplicitous journalist or producer, the scientist isn't exactly as blameless as a newborn babe either: Exactly how much vetting, checking, and referencing did the scientist do before the interview? Aren't we trained to ask critical questions, not just blindly believe what we're told? Would a five-minute internet search have revealed everything the scientist would need to know?

Sometimes it's out of laziness. The scientist does the interview, then moves on with their life, forgetting it even happened. They don't know and don't care what happens after that. And because they don't care, they don't spend an ounce of effort to actually effectively communicate to the journalist, let alone the public. They don't clearly explain their jargon, they don't take a high-level view. They care more about accuracy of details (the way you would write a journal article) than achieving understanding.

And sometimes it's very much intentional. As in, *on purpose*. It's a scientist wanting, desiring, craving the public spotlight, no matter the cost to themselves, their career, or the entire field of science. The kind willing to not only talk about bad science but *do* bad science. To publish rubbish articles in junk journals. To talk about fringe ideas as if they have a shot at legitimacy. To advance their own specious claims in the public sphere, because the vast majority of their colleagues have already rejected them and it's the only stage left.

I'll admit it can feel really, really good to talk to the media. To have a journalist seeking your wisdom and insight. To sit down in front of the camera. To see your name in print. And that can be a lure for many scientists, who do not have the training or expertise of, say, an experienced politician. The scientist who can simply be taken advantage of, and ironically participate actively in that process the whole time.

One could make the argument that sexy, exciting news gets people interested in and engaged with science, but that's pretty weak. Surely there are ways other than misrepresenting—and outright *lying*—about the fundamental nature of the scientific discipline. Really, *that's* the way to bring people in?

It's a nasty feedback loop: scientists, usually cautious, try to be as careful as possible in their papers. But once the media gets a sniff of a good story, they loosen up in interviews. That attention brings a lot of positive notice, from both the public and their peers. They're the center of attention. They're *important*. They have things to say. Their papers get more citations. Their university or lab gets more visibility.

They try even harder next time. Their younger colleagues look up to them for an example to follow. And it all starts again.

So once again I ask why. How did we end up in such a deplorable state? Why are we embarrassed by science media instead of celebrating it? Why are journalists doing all the work and not scientists themselves?

It isn't better because nobody wants it to be better—from their perspective, everything works A-OK. Universities and labs get tons of free exposure. News outlets and producers get tons of eyeballs and clicks. And scientists get tons of "science" communicated to the public (assuming they even care).

The scientists don't want to fix it because they don't want to bother. It's too tough, too complicated, to do the job themselves. And when they do talk to journalists, they're under no pressure to try to be good at it. They're busy, and there are no incentives to change it.

And the more the machine crunches on, distilling and twisting and distorting science, we all suffer for it.

That's all depressingly horrible, but traditional media isn't the only avenue available for scientists to connect to the general public. Sadly, the other outlets don't capture nearly the number of eyeballs and interest as the media.

There's formal science education—the kind you learned in school. The state of modern science education is . . . sigh, an entirely new book, so I'll leave that as a homework exercise for the reader.[9]

When a scientist visits a school, it's a great experience for all involved,[10] but it doesn't happen nearly often enough. You can guess the reason why: it begins with "b" and ends with "usy." The same goes for all those wonderfully creative public outreach events, which strangely seem to be mostly organized by young scientists—the ones with few career prospects in the first place. The more entrenched a scientist becomes, the more they believe, sadly, that science outreach "isn't for them."[11]

As for social media (and by this term I mean all things digital), that's a strange beast indeed. Many scientists take to blogs and social posts to talk about their work, their role in society, and thoughts on science in general. But although "many" sounds like a lot, it is a pathetically small percentage of the entire scientific population, and only rarely do they enter the social sphere with the intent of communicating science.[12] This practice is a double-edged sword: a scientist can build a network of followers who look to them as a trusted source of information,[13] but due to the fractured reality of the social landscape, they end up speaking to their fans (or at best a certain segment of the population) . . . and nobody else.

This occurs in real life as well. I don't know how many talks I've given to science clubs and university-organized events that were to rooms full of . . . middle-aged white guys who were already fans of science. It is extremely difficult to draw new audiences into science presentations, and yet scientists continue to follow the same molds made decades ago in a completely different era: school talks and monthly seminars.

When it comes to YouTube and TikTok, where a tremendous percentage of young people consume content at a frightening rate, the vast majority of science presented on those platforms is (a) not presented by actual scientists or (b) not actually science at all.

For example, the top YouTube science shows, like *Vsauce* and *SciShow*, are not hosted by scientists but by entertainers. Just as with traditional media, that's not a bad thing in and of itself, but it forces me to ask: where exactly *are* the scientists? No, it doesn't take a PhD in a field in order to be able to effectively communicate it to the public, but wouldn't it be nice if someone deeply involved in their discipline—to the point where they independently advanced new knowledge—could share what they know and love? Why are we leaving it to anybody else?

Adding to that is the noxious sewage pipe that is the "recommended videos" algorithm. There are so many videos on those platforms, producing so many raw hours of content, that it's almost impossible to get any visibility unless you speak about certain hot topics.

The average viewer encountering the average YouTube or TikTok page, wanting to watch a cool science video or two, will be presented with a buffet of nonsense videos—all of them looking and sounding legit. For example, typing "astronomy" into the little search box on YouTube gave me (your results may vary) a legit video hosted by a scientist (yay!), a couple documentaries on the universe that are almost entirely wrong, a video about ten discoveries that "scare" astronomers, a video on the "biggest mystery in astronomy" that isn't, and so on, with absolutely no way to discern which are legit and which are not.

If the state of science communication is broken, why are scientists seemingly unwilling to change it? I'll tell it to you straight: scientists who do decide to engage more with the public risk ridicule from their peers. If they're young, then they can largely get away with it, as long as they don't devote too much time to it (and I have personally encountered faculty members who have explicitly forbade their students from engaging in outreach). But once they're on the tenure track, then anything that doesn't directly advance their careers is implicitly verboten. And if they become a celebrity? Forget about it. The celebrated—and popular—astronomer Carl Sagan was denied tenure at Harvard and denied nomination to the Academy of Sciences, despite having a respectable academic career—the reason is largely thought to be the disdain of his popular presence. I would say, "Oh, the '80s were a crazy time," but it's just as bad today.[14]

The truth is there's absolutely no *incentive* for scientists to engage in outreach and communication. Sure, it's "celebrated" in a weak way, with half-hearted campaigns by universities to encourage scientists to connect with the public more (especially through the lucrative press release), but scientists in academic and lab circles know the hard truth: at the very best, science outreach is seen as neutral, and more often it's a liability.

When applying for prestigious fellowships and graduate programs, a certain measure of science outreach on your résumé is seen as nice. Just . . . nice. You better have the academic chops to prove your worth—and that's exactly what the committee will look into—but a few community events here and there will demonstrate that you're not a mathematics-generating robot and actually have some soul to go along with your flesh and bones. But too much? Well then *obviously* you're not interested in a career in science; otherwise you wouldn't be wasting your time not doing research, right?

At the higher levels in the science career path, as we've seen, outreach and communication are simply not an option. They will be scored negatively against you: remember, the primary job of the modern scientist is to win grant money. Having a successful outreach program (or, heaven forbid, a *book*) does not bring home the grant bacon, and if you're not obsessively focused on that singular task, then exactly what are you here for?[15]

Because scientists aren't *compelled* to do outreach, as a part of their duties, they simply don't do it, unless it's out of the kindness of their hearts.

Here's the real irony. Grants funded by the National Science Foundation must ostensibly spend a portion of their awarded time and money on some form of science outreach or communication. It's true! It's called "Broader Impacts" and must be a part of every single proposal.[16] On the surface, it looks like this problem has already been solved by our national funding agencies. If you want public money, the public must benefit in some way.

Surprise, surprise: it doesn't work.

For one, "Broader Impacts" can be interpreted . . . well, broadly. As long as you can spin that your project has some benefit for society as a whole, you're golden (I've written research proposals that somehow connect studies of the large-scale structure of the universe to curing the ills that plague our nation). That's assuming that you even bother to address the question—many awarded grants don't have much of a Broader Impacts section in the first place.[17]

And also, *you don't really have to do it.* You can outline all the grand plans and schemes that you want, allocating resources for workshops and teacher engagements and museum exhibits, and if you don't manage to get around to it . . . well, it's a slight knock for a renewal but not a deal-breaker. Nobody is following up. Nobody is knocking on your door. Nobody seriously penalizes a successful researcher for not following through with their plans. *Nobody cares.*

Still, some lip service must be paid to the people holding the purse strings, and so sometimes a grant-winning researcher with some Broader Impacts requirement will turn to their postdocs and graduate students to "do that outreach thing." Hence: younger scientists quasi-compelled to do it, reducing their research output and putting them at a competitive disadvantage, because the senior researcher wouldn't dare do it themselves.

In short, despite the name, Broader Impacts don't really make much of an impact.[18] It's a cynical farce.

While individual scientists may participate in outreach and communication, the community as a whole is not invested in connecting with the public at all. There is no incentive structure, no reward, for a scientist to engage with the public. There's nothing in it for them.

It's easy enough to blame "the system" for all the broken aspects of our society, but as we've seen time and time again in these chapters, "the system" is made of people, and what "the system" does is what the people involved want it to do. Scientists aren't incentivized to connect with the public because they don't want to be incentivized. They want an excuse to not have to do it. They want to focus on their research, sitting comfortably in their offices or uncomfortably out in the field, without being held accountable to the people paying their checks.

It's laziness and apathy. *Scientists do science.* That is their job. That is their position. That is their focus. Their goal is to understand the world around us and communicate that to their peers in order to achieve recognition and reward for their efforts. Things like teaching and committee service are seen as a necessary part of the academic life—but if they're flush with enough grant money, they can buy themselves out of some of those duties. But outreach? That's not what a scientist does. That's not the reason a scientist exists. That's not the point.

What's the end result of all this refusal to communicate with the public in any meaningful, sustained, committed way?

A public that is becoming more and more distrustful of scientists and disconnected from their work with each passing day. A public that is left behind. Yes, scientists enjoy a great degree of trust in modern society—but that trust is fragile.[19] And as the years go by with the same broken structures perpetuated, scientists continue to demand trust instead of earning it.

And while trust in scientists is relatively high on average (73 percent of US adults think that science has a "mostly positive" impact on society), that number is far lower for Blacks (59 percent) and adults with low knowledge of science (54 percent).[20] Minority audiences often get left behind when it comes to science outreach, which entrenches and enhances their existing antiscience sentiments.[21]

Hmmm, trust in science is highest among the groups of people who are already inclined to be the recipients of science communication. Way to preach to the choir, I suppose, but aren't we trying to get *all* of society onboard with this "science is awesome" thing? We openly wonder why certain groups and populations don't trust science . . . and we never bother to call them on the phone.

And won't somebody think of the children? Those same kids we are begging to go into STEM fields rarely encounter a living, breathing scientist. And without that direct human connection, they have a very rare chance of realizing that a career in science is actually for them. Without a role model, the kids we are raising are left rudderless. The end result: while younger kids *love* science, by the time high schoolers are considering college degree and career options, science is uniformly near the bottom of the list.[22]

The lack of science communication crystallizes stereotypes. The less the public—especially kids—see and interact with scientists, the more they see them as crazy-haired, lab-coated caricatures, instead of what they actually are, which is just . . . people.[23] Smart, motivated, people, but people nonetheless. If we want people to realize the truth and have a more accurate image of "scientist" pop into their head when they hear the word, *maybe we should introduce ourselves.*

Those stereotypes work in reverse too. The more scientists engage with the public, perhaps the more they will take off their ivory-colored glasses and remember that normal people are just that: normal people. Not stupid, not ignorant, not hateful and out to destroy science. Just people, living their lives, working their jobs, raising their kids. Trying to get by in a difficult, complicated world. A public that can be energized, excited, and informed by science—if they feel like they're being spoken to as equals and collaborators, not as fools awaiting instructions.

It's depressingly ironic. By continuing to ignore the public, scientists are reinforcing the negative views that they hate so much: that science is corrupt, complex, heretical, and limited.[24] Scientists want to be seen as leaders, informers, and guiders, but meet constant resistance—resistance that often takes the shape of those four words that I just typed. And until they solve that underlying trust issue, their role in society—while respected by many—will never meet its full potential.

And then there's the whole pseudoscience thing.

I talked earlier about how the average member of the public has a really hard time distinguishing good science from total garbage—heck, sometimes even I have to do a double or triple take to make sure that what I'm reading is legit. This blurring of the lines between legitimate and illegitimate sources of information (notice that I didn't say "scientists" because, as we've seen, even scientists can be sources of junk knowledge), caused by the unwillingness of scientists to actually be present and relevant in the everyday lives of everyday people, directly causes the rise and proliferation of perhaps the greatest shadow to haunt science in modern society: pseudoscience.[25]

Now for the purposes of illuminating this discussion I'm going to put a lot of related phenomena under the broad umbrella of "pseudoscience." For starters, you can take all those "ghost hunter" TV shows that look Very Serious but are actually complete farces. Scientists, as a community, do not believe in the existence of ghosts. Or put another way, whatever a random individual scientist may believe on the topic of souls or afterlife, no scientist believes that "ghosts" as we commonly understand them have a measurable presence on the material world—they are, by all definitions, supernatural. But ghost hunters make it look like science cares. They carry around gadgets and gear to make measurements, so they've acquired the trappings of scientists (sometimes they even literally wear lab coats, even when they're not hunting lab ghosts), but they lack the rigor that actually makes a scientist a scientist: rooting out bias, making testable hypotheses, and a willingness to be wrong.

They look like they're "doing science" but it's all dishonest and twisted.

Related to the ghost-hunter style of pseudoscience is the crackpot—the loner with an incomprehensible theory to explain the pyramids, dark matter, consciousness, whatever. To a trained scientist, they're spouting pure nonsense. They have all the right jargon words, but in jumbled order with no logical connection. Imagine taking the sentences of these chapters and running it through a blender. All the right words and phrases make an appearance but with no cohesive sense—it might as well be gibberish (and I'm sincerely hoping that my intentional writing rises slightly above that bar).

And then there's the denialism and outright antiscience elements found in all societies, like the ones we've already explored. There are large and noticeable (and loud) segments of the population that believe that the Earth is not getting warmer due to human activities. There are people who do not understand or trust the process of genetically modifying crops. There are people who refuse to get vaccinated. There are people who *literally believe that the Earth is flat.*

And there's so much more.[26]

All these antiscience groups publish articles, podcasts, YouTube videos. They host organized meetings. They advocate in the public space for their views (yes, even the flat-Earthers). And what they create has all the frills of science. It looks like science. It smells like science. It seems as authoritative as science.

And lastly, there are the fakers. The people pretending to be scientists, with no qualifications, no credentials, no knowledge, no expertise. But they're perfectly willing to talk about science to the public, guide informed policy decisions, and become a source of trusted information. And they don't care if they're getting it right or not. They care about the fame, notoriety, and money (you know, all the things that usually come with a career in science, I type as I roll my eyes). They don't discriminate in their sources (or do, based on the highest bidder). Their authority doesn't flow from a position of experience but rather through a suitably slick presentation. If science had heretics, these would be them.

Scientists have watched over the decades as all these forms of pseudoscience have flourished, grown, and become entrenched, cutting across political, religious, and social divisions with ease. Pseudoscience has taken the public's trust in science and aimed it in their nefarious direction. They're perfectly willing to openly tarnish the name and practice of science for their own gain.

For every awesome chock-full-o'-science video on YouTube, there are a dozen horrible ones, often user generated and shockingly easy to find.[27]

For every determined science communicator on the internet, there is a tsunami of information. A howling hurricane of garbage compared to the whisper of genuine knowledge.[28]

For every climate scientist diligently trying to inform the world of the potential harm of our actions, there is an army of denialists drowning out the discussion.[29]

For every doctor and epidemiologist who tried to guide our world through the worst pandemic in a century, the scourge of pseudoscience claimed more and more lives of our loved ones.[30]

Pseudoscience causes measurable, demonstrable harm. To the institution of science. To the public. To you. And yet it shows no sign of stopping, or even slowing. If science truly had a foe to grapple with, it's not religion or some political affiliation; it's pseudoscience in all its hideous forms.

This pseudoscience spreads for its own seemingly illogical reasons. Scientists bemoan its presence and fight irrationality with reason, made-up fantasies with facts. And yet it never works. Pseudoscience grows not because people hunger for an objective representation of reality but for a myriad of motives: identity with a group, a distrust of authority, a desire to make sense of a senseless world, and more.[31] Even something like vaccinations, which seem like a simple matter of do-you-want-your-kids-to-get-sick-or-not, meets resistance for a complex intersection of reasons; it's not just a simple matter of "people are antiscience."[32]

And as pseudoscience grows, it becomes harder and harder for the general public to tell the difference,[33] and outright denial of science (and everything that the scientific worldview represents) grows in lockstep.[34]

All these forms of pseudoscience (literally, *false-science*) have the same origin point: the distressing lack of actual, real, human scientists in the public sphere. Hiding in their offices and labs, every single time that a scientist refuses to engage and connect with the public, they make the problem worse.

Scientists *want* to blame the media for all the twisted headlines and junk TV shows (and they're not exactly wrong—journalists, editors, and producers aren't blameless).[35] They *want* to blame the crackpot and the ghost hunter and crackpot (and they're not exactly wrong—some people are perfectly willing to abuse any authoritative institution for personal gain). They *want* to blame their colleagues for screwing it all up again (and they're not exactly wrong—there are definitely some devious and just plain *bad* scientists out there in the world).

The *want* to blame the public, for not listening to them when they do speak (and they're not exactly wrong—many people, for many reasons, want to reject science either in part or in whole).

But they don't want to blame themselves, because blaming themselves means they would have to shoulder the burden of trying to fix it.

Why do we constantly expect the general public to understand science to such a high degree of sophistication? Why do we expect them to simply *know* the difference between science and pseudoscience? Do we ever expect or demand this level of expertise when it comes to medicine, law . . . car repair?

Scientists assume that since public trust in their institution is relatively high, that tackling pseudoscience is somebody else's problem. Patting themselves on the back, they see no need to take action (let alone try to figure out *what* action might be the best to take). But "trust" is not the whole story: there's also credibility. When times get tough, people may trust science all they want, but we have no guarantees that they will choose their "science" sources correctly—they're just as like to mistake the pseudo kind for the real deal.[36]

Yes, science journalism and sensationalism is out of control. Yes, the digital landscape is mostly a swamp of user-generated misinformation. Yes, there are many hardworking scientists out there doing their best to fight the good fight.

And it's not working. As trust in science declines, as pseudoscience rises, as scientists continue to maintain their distance from the problem, the ultimate victim of this vicious cycle is science itself. Modern science is not funded by a handful of ultrawealthy benefactors, like it was in its infancy. Science is paid for by the people, by the public. They're the ones footing the bill. But if credibility erodes and people don't know where to turn to for trusted information, where do you think they'll spend their dollars?

I know where I would.

In many ways, the people have already spoken. Science funding continues to decrease year after year. Despite all that supposed "trust," nobody is willing to actually pay for science to continue anymore. With diminished and limited access to scientists, the public turns to anyone who sounds convincing enough. I'm not trying to portray John and Jane Q. Public as ignoramuses; I believe that the average person is wonderfully intelligent, but I do not pretend that they'll know "real" science when they see it, because they have so few good examples. In return, scientists become even more defensive and entrenched, refusing to engage with a hostile public. Spurned, the public loses sympathy and excitement for science, reducing financial support . . .

The divide between scientists and the public grows from a minor crack to a tremendous gulf. A Grand Canyon of intellectual distance.

Scientists—real, actual scientists—lose respect and prestige. Minority groups who already have trouble trusting scientists get their worst fears confirmed. Fewer and fewer students see STEM as a viable and interesting career path.

We all lose.

The sad thing is science communication is powerful. It's enlightening. It's fun. It's informative. It's *therapy*—it creates a space where people can feel safe enough to ask questions and follow their curiosity, a time where they can mimic the scientists they admire and swim in a sea of ignorance—but filled to the brim with boundless possibility. It's one of the most poignant aspects of science: the joy of not knowing the answer ahead of time and letting nature take you by the hand and guide you.

It's a shame the public doesn't get to experience that sublime joy as much as they could.

I'm not saying that we can cure all of society's ills, make everybody happy, and solve every problem plaguing modern science by opening our doors just a little bit more. But . . . it can't *hurt*, right? Is there any evidence or example where open, honest discussion and exchange of ideas led to all parties involved departing less enriched than when they entered?

And I'm not saying that every scientist is guilty of closing themselves off from the public, locking themselves in their monastery—I mean, university—and refusing to engage with an increasingly frustrated public. There are many, many wonderful science communicators out there, representing all fields of science. But they are such a small fraction of the total science population—and especially the total amount of potential intellectual capacity that scientists can provide—that the meager scraps that the public do get are indistinguishable from nothing at all.

How do we begin to heal these wounds? How do we unwind the damage that's already been done? How do we deconstruct the stereotypes that scientists hold of the public? How do we turn arrogance and superiority into an attitude of public service?

Well, we could always *make* scientists do it.

The public pays for science. The vast, vast majority of modern science conducted around the globe is at the goodwill and good money of the taxpayer. I firmly believe that the knowledge that science creates does not belong to scientists. I firmly believe that scientists do not work to increase the knowledge of scientists but to increase the knowledge of *humanity*.

Scientists owe it to the public. To humanity. To you. You paid for it, with your hard-earned money. Scientists work at the pleasure of the public. They are servants, not masters. They are creators of knowledge, not hoarders. You *deserve* to have the output of science made available to you.

I believe that it is the moral and ethical duty of every scientist to make their work accessible to the public. If the only people who can understand the work of a scientist are the few working on related problems, then that's simply not good enough.

We can get serious about Broader Impacts. We can make scientists take, I don't know, a *public communications for dummies* course as a part of their undergraduate or graduate education. If "translating" scientific jargon and methodology into some form that can be understood is too difficult for the average scientist (as they often claim), then we're not doing a good enough job. Scientists have to pick up a dozen skills and more in order to be successful—what's one more?

We can insist that universities and departments revise—if not outright discard—their antiquated selection and promotion guidelines. If faculty had to engage in a certain amount of outreach as a part of their tenure qualifications, if they were actually *rewarded* for outreach, then you can bet your bottom dollar they would get it done. And probably surprise everyone with how well they do it. Remember, scientists already know how to communicate effectively. They do it every single day. What I'm proposing isn't much different from their existing day-to-day.

Would it all be perfect? Would it all be stellar examples of elucidating our complex, mysterious universe for lay audiences? Absolutely not. The more important question is this: what can we start doing *right now*? Right now we're facing a problem of *volume*, not finesse. Once we have more actual scientists doing actual communication with actual people, we can start to worry about the finer points and maximizing the effectiveness of their communication efforts. The more scientists break down walls of jargon, letting the light of legitimacy in, the more pseudoscience will have to hide in the corners.

Scientists should be *everywhere*. They should be on TV. They should be in the classroom. They should be at the corner bar. Scientists ought to spend a fraction of their professional time away from the lab, away from the office. They should be talking.

Will this reduce scientific output? Of course. Less time spent on hardcore research means science will advance less. But as we saw earlier in these pages, science isn't really "advancing" in the way that you and I would naturally expect "advance" to mean. Would we really miss anything significant about the universe if every scientist had to spend 10 percent of their time—let's call it a scientific tithe just to be funny—engaging with the public?

Throughout these chapters, I've proposed sometimes radical solutions for the direst of problems facing modern science. All those solutions have focused on scientists themselves because I believe they are the principal agents behind those problems. But in this case, I have a radical suggestion for you, the reader: stop reading science news stories. Don't reward those headlines with your clicks and your precious eyeballs. Don't feed the machine. Stop paying attention to hyped-up science news. I have a saying, which some of my fans have dubbed "Sutter's Razor": if it's interesting, it's probably wrong.

If you see a headline that seems too exciting to be true, why bother wasting the time and energy to read it?

Instead, I suggest you follow scientists themselves. For whatever field you're a fan of, whether it's astronomy or anatomy, there is an array of scientists on the internet discussing their research and the research of their communities. Seek them out, on social media, on blogs, on YouTube, anywhere. Follow them. Ask them questions. Be delighted by their responses. They may not be perfect. They may not be polished. That may not always make sense. But they'll always, always be real.

If we do this enough, the media will catch our drift: we're simply not interested in science news and discussions unless it's from a trusted source of *our* choosing.

But besides that, we also need to educate scientists. All too often they believe in the so-called deficit model, which posits that the primary reason people believe pseudoscience rather than actual science is ignorance—people simply don't know any better, and a flood of information will purify their minds.[37] This . . . doesn't work. Convincing the public of the veracity and power of science requires understanding the intertwined roles of faith, ideology, and background.[38] Scientists can't just lazily copy their journal articles to a web page and call it a day: they have to take into account the needs and wants of their audiences, which are very different than the wants and needs of their colleagues (which mostly consist of wanting specific questions answered, like the statistics, methodology, and location of the coffee maker). In the end, effectively sharing science requires empathy, trust, honesty, and respect, even when confronting a room full of antagonists and nonbelievers.[39]

A better name for the future of science communication might be "science connection."

Just as scientists want to arm the public to filter out bad science media, we need to arm scientists to do the same. We need to train them on how to speak to journalist, on how to understand the goals and motives of a media creator (which are, again, different than the goals of their colleagues). We need our scientists to insist on review of articles prior to publishing, to ensure that their quotes are appropriate and taken in proper context. I know that this is not the usual practice of any kind of journalism, but the usual practice of journalism is not serving the needs of science or the public very well, and so we need to discard it. The less scientists are willing to play this game, the more journalists will have to adapt.

We also need to emphasize in science education the human aspect of this grand question of knowledge. We need to instruct future scientists that the best leaders are servants of the greater good. We need to teach them that their talents and skills don't separate them from humanity but weave them deeply into society. We need a generation of humble and patient scientists who celebrate their role in humanity's quest to make sense of the world around them—and that it's something that *all* humanity takes part in, not just them. We need scientists to embrace the general public not as a source of potential funds or an ignorant mass but as a pool of excited, eager, equal learners. A population *craving* more knowledge of this world, more than ready and more than willing to support, defend, and celebrate science.

We need scientists to understand that effective science communication requires far less work than they think it does. A more active presence on social media. An outreach talk here and there. An interview. A friendly chat with friends and family. A small fraction of their time and talents devoted to a purpose outside academia. With the right incentives in place, the right attitudes taught, and the right dose of humility, science communication, outreach, and sharing will become the natural consequence of a scientific career, not a bothersome distraction. And if we have to drag some senior researchers into the twenty-first century kicking and screaming to do it, so be it.

Science outreach and communication isn't about big-budget TV shows or hard-hitting news articles. It's about humans connecting with humans, sharing their passions and their knowledge, spreading enthusiasm for the wonders of the natural world, the wisdom brought by the scientific method, and the sublime joy of discovery.

After all, we're all in this voyage together.

Epilogue

The Virtues of Science

WELL, FOLKS, THAT WAS ONE HECK OF A TRIP, WASN'T IT? Like I said at the beginning, of course this book is going to make science look bad, because this is a book about the bad parts of science.

We've seen how sick modern science is. We've seen how very human and very real vices have found their way into the scientific enterprise and corrupted what should be among the most noble and pure of human endeavors. Science in these early decades of the twenty-first century is in pretty rough shape, and I fear the worst.

Perhaps you started this book with aspirations of a career in science in mind and reached this point wondering if a PhD is worth it. I have a couple things to say to you. First, a PhD in science is a very powerful, liberating thing: you're almost guaranteed to get a good job, somewhere (just maybe not in academia), doing something really cool. And the adventures that you have in science, no matter how long your career lasts, are unlike anything else.

In my original draft of this book, I organized the chapters around the seven deadly sins of envy, sloth, gluttony, lust, wrath, greed, and pride (as a way of communicating that the issues with modern science practice are born from the same afflictions that permeate any human institution). So maybe someday I'll write a sequel, *A Success in Science*, and use the seven virtues (the flipside of the deadly sins) as a guide. Here's a preview:

Science is awesome, folks. Through the tools of the scientific method, and the hard work of generations of scientists, we've expanded the frontiers of human knowledge more than we could ever dream possible. We've peered into the first moments of the very existence of the universe. We've explained the variety of species on our planet. We've listened to the secrets whispered by billions of years of the ground beneath our feet.

We've organized, categorized, listed, sorted, and *understood* like never before.

The knowledge we've gained from science has made us commanders of our environment, for good or ill. We've harnessed subatomic forces and the light of the sun itself. We've reshaped and resculpted the face of our planet. We've conquered diseases that have scourged humanity for millennia. We've sent our robotic emissaries beyond the bounds of the solar system.

If the Earth were to be obliterated tomorrow, our memory would endure.

We've turned the universe into the playground of our curiosity. We've reveled in the pursuit of knowledge, the gaining of new wisdom. The world we see around us is more colorful, more vivid, more *alive* than the one of our ancestors. It's a world not ruled by capricious spirits and mysterious forces but governed by natural laws and harmonious rhythms.

Science is beautiful.

And being a scientist is really, really fun. I cherish the memories I made working late nights as a grad student, trying to get some piece of code to work in my attempts to simulate some of the most gargantuan events imaginable. I remember fondly the work I did in my postdoctoral positions as we came up with new ways to understand the voids of our cosmos. Every paper represented a mountain of effort, but it was worth it.

I got to be friends with some amazing, smart, talented, sharp, witty people. I got to chat with some of the smartest and most engaging brains our species has ever produced. I got to hear new ideas, new results, new discoveries when they were barely sketched out, long before the public ever became aware of them.

Every time I got a new result, when the simulation finished or the analysis completed, I would stare at a simple plot or graph, a modest summary of an enormous amount of data. And sometimes I would do just that—stare—for a good long while. For in those moments, I felt myself on the edge of discovery, an explorer witnessing a new vista for the first time. For a brief moment, I was the only human being in the entire world, and indeed throughout history, to have that knowledge.

It's a thrill that little can match.

Science is a love of curiosity, a love of learning, a love of discovery. The methods of science are there to sharpen and focus our intellect to laser-like precision, allowing us to concentrate on useful, answerable questions and deliver useful, powerful results. We are all curious, every one of us, to some degree, and science is one of the most powerful forces for engaging that curiosity.

Science is full of virtues, especially the top seven.

Science is prudent. Scientists emphasize the use of reason, wisdom, insight, and knowledge to reach decisions. There may be lots of negative stereotypes of scientists, but there are positive ones too. Scientists are the ones you turn to for trusted, honest advice. The best kind.

Science is just. The organization of science is founded on beautiful, strong principles. The highest accolades are supposed to go to the ones who earn them through merit. Recognition is based not on creed or color but achievement and contribution. Arguments can be waged, but evidence is the ultimate arbiter—eventually, all scientific disputes do come to an end, one way or another.

Science is temperate. We don't blindly rush to any conclusion just because we want it to be true. We test, and test again. We experiment. We place controls on our research to filter out our own biases and preconceived notions. We question our assumptions. What we ultimately believe is weighted by the evidence, regardless of what we personally wish to be true.

Science is courageous. Facing the evidence regardless of beliefs is *hard*. It's against human nature to see a result contrary to your desires—and even more so to publish and promote it anyway. Changing your mind is one of the most difficult things you can do, and the very process of science is designed to *make* you change your mind. Being a scientist is living in a constant state of ignorance, where any possibility can be accepted or discarded based on the evidence, where prevailing wisdom can come clattering to the floor overnight, where everything you've ever worked for can be for nothing. And yes, speaking about science can be hard. Not everyone is a fan or supporter of the methods or results that we obtain.

Sometimes people really do hate scientists and what they represent—it takes real guts to stand up to that.

Science is faithful. Ironically. Scientists believe that the scientific method and process actually produces meaningful results. It's this belief, this fundamental philosophy, this faith, that propels scientists every day to get in their labs or offices or out into the field and keep grinding away. Science is far, far from easy, and to get out of bed in the morning requires a deep commitment and passion that eventually it will all be *worth it*. That you'll get that result, you'll gain that understanding, you'll achieve that insight.

Science is hopeful. The vast majority of scientists believe in a good world and a better future. They want to be lights to humanity. They hope that they'll learn more than their predecessors did and that they'll be able to pass on to the next generation a new foundation of knowledge. They hope that the world will forge that knowledge to good, noble purposes.

And lastly, science is charitable—after all, most scientists certainly aren't in it for the money. For their skills and expertise, a typical PhD-trained scientist could command roughly double their salary if they ventured outside academia, and all those years spent working on their PhD set them behind their nonacademic peers in terms of salary potential and life savings. People go into science to fulfill a passion for curiosity and new knowledge, not for the mountains of cash. Science has always been, and likely always will be, a service to humanity.

That would be a wonderful book to write! But not today.

I'm proud to call myself a scientist. I'm excited for what scientists will uncover about our universe for years to come.

I suspect that this book will make some (many?) people upset. In a way, I hope it does—the point of this book is to shake things up a bit. I look forward to the conversations and debates to come (but not the YouTube comments). Like I said I would at the beginning, I've used some pretty tough language to describe my colleagues. If you are a scientist and feel personally offended by my words . . . well then, maybe we should explore why you feel offended. Let's talk.

Science is sick. It's fallen into the same traps that human institutions have fallen into since time immemorial.

The modern institution of science creates a toxic, overcompetitive atmosphere where young scientists attempt to race to the top, only to find that they're on a hamster wheel going nowhere fast. Scientists commit and enable fraud because publications count above all else—as a part of that same toxic, overcompetitive atmosphere. In the light of reduced funding from national agencies, scientists don't take risks but favor incremental improvements that are easy to quantify. In light of that same reduced funding, scientists choose to close themselves off to the public and disengage from the very people paying their bills. Scientists seek influence in the political realm while trying to distance themselves from it. They pretend to meritocracy and use that as an excuse to maintain and perpetuate racial privileges.

As a result of all this, the public continues to trust scientists less and less. Young scientists find fewer and fewer career options open to them. The future of science grows ever more uncertain.

But when confronted with our own shortcomings, our challenge is not to give in to those vices but to rise above them and become better versions of ourselves.

And I believe that it's not too late to rescue science from its shortcomings. The problems facing modern science aren't anything especially new to the human condition, and it's precisely because they're nothing new that we have the hope to make science better.

It might take risky, bold action. We might have to eliminate peer review and the tenure system. We might have to shake up, or outright dismantle, our higher education system. We might have to ostracize our highest-performing scientists for abusive or fraudulent behavior. We might have to sacrifice some speed in gaining new knowledge in order to spend that time engaging with the public and leaders.

We might have to make some decisions when we're not guaranteed success.

I don't know if any of my recommendations will work. But what should worry science about a little experimentation? Shouldn't we take our vaunted tools and methods and apply them to our very own institutions? Shouldn't we be willing and eager to critically evaluate *how* science is done and let the evidence tell us if we're doing right by society and ourselves?

Come on, scientists, let's shake things up a bit and see what happens. If I'm wrong, then I'm wrong—that's how science works. But if I'm right . . .

I hope you have a new perspective on science after reading these pages. I hope you understand why scientists may be distrusted—why you yourself may distrust science—and what we can do about it. I hope you see the value in preserving this beautiful institution that we've constructed and support it just as strongly as you advocate for reform.

I truly want all people to see the way that I, and my colleagues, see the world. Through a lens of objectivity, reason, and beauty. I want people to understand, appreciate, and value science. I want scientists to be respected and admired. I want science to thrive for generations to come.

I wrote this book because I want our kids and their kids to be able to strive boldly into the unknown, past the edge of human knowledge, assured that society will support them in their quest for understanding and welcome them back warmly when they return. It's not too late to make that happen, but we have work to do.

Notes

Chapter 1

1. *Astronomical Journal*, "Page Charges," https://iopscience.iop.org/journal/1538-3881/page/Page%20charges.

2. R. Johnson, A. Watkinson, and M. Mabe, *The STM Report: An Overview of Scientific and Scholarly Publishing*, 5th ed. (The Hague, Netherlands: International Association of Scientific, Technical and Medical Publishers, 2018), 5.

3. Ibid.

4. C. A. Chapman et al., "Games Academics Play and Their Consequences: How Authorship, *h*-Index and Journal Impact Factors Are Shaping the Future of Academia," *Proceedings of the Royal Society B: Biological Sciences* 286 (2019): 1–9.

5. Charles Gross, "Disgrace: On Marc Hauser," *Nation*, December 21, 2011, https://www.thenation.com/article/archive/disgrace-marc-hauser/.

6. L. Bornmann and H.-D. Daniel, "What Do We Know about the *h* Index?," *Journal of the American Society for Information Science and Technology* 58, no. 9 (2007): 1381–85.

7. Chapman et al., "Games Academics Play."

8. Gross, "Disgrace."

9. Nicholas Wade, "Harvard Finds Marc Hauser Guilty of Scientific Misconduct," *New York Times*, August 21, 2010, https://www.nytimes.com/2010/08/21/education/21harvard.html.

10. T. Mahmood and P. C. Yang, "Western Blot: Technique, Theory, and Trouble Shooting," *North American Journal of Medical Sciences* 4, no. 9 (2012): 429–34.

11. James Glanz and Agustin Armendariz, "Years of Ethics Charges, but Star Cancer Researcher Gets a Pass," *New York Times*, March 8, 2017, https://www.nytimes.com/2017/03/08/science/cancer-carlo-croce.html.

12. "Carlo Croce Loses a Round in Legal Bid to Be Reinstated as Dep't Chair," Retraction Watch, February 4, 2019, https://retractionwatch.com/2019/02/04 carlo-croce-loses-a-round-in-legal-bid-to-be-reinstated-as-dept-chair/.

13. D. Fanelli, "How Many Scientists Fabricate and Falsify Research? A Systematic Review and Meta-analysis of Survey Data," *PLOS One* 4, no. 5 (2009): e5738, https://doi.org/10.1371/journal.pone.0005738.

14. B. Martin, "Scientific Fraud and the Power Structure of Science," *Prometheus* 10, no. 1 (1992): 83–98.

15. Fanelli, "How Many Scientists Fabricate and Falsify Research?"

16. Ibid.

17. P. A. R. Ade et al., "Detection of *B*-Mode Polarization at Degree Angular Scales by BICEP2," *Physical Review Letters* 112, no. 24 (2014): 241101.

18. P. A. R. Ade, et al. "Joint Analysis of BICEP2/Keck Array and Planck Data," *Physical Review Lettetters* 114 (2015): 1–17.

19. Steve Mirsky, "No, No Nobel: How to Lose the Prize," *Scientific American*, May 19, 2020, https://www.scientificamerican.com/podcast/episode/no-no-nobel-how-to-lose-the-prize/.

20. M. L. Head, L. Holman, R. Lanfear, A. T. Kahn, and M. D. Jennions, "The Extent and Consequences of *P*-Hacking in Science," *PLOS Biology* 13, no. 3 (2015): e1002106, https://doi.org/10.1371/journal.pbio.1002106.

21. R. Walker and P. R. da Silva, "Emerging Trends in Peer Review—a Survey," *Frontiers in Neuroscieince* 9 (2015), https://doi.org/10.3389/fnins.2015.00169.

22. W. Stroebe, T. Postmes, and R. Spears, "Scientific Misconduct and the Myth of Self-Correction in Science," *Perspectives on Psychological Science* 7, no. 6 (2012): 670–88.

23. Chapman et al., "Games Academics Play."

24. D. M. Markowitz and J. T. Hancock, "Linguistic Obfuscation in Fraudulent Science," *Journal of Languages and Social Psychology* 35, no. 4 (2016): 435–45.

25. See Retraction Watch, https://retractionwatch.com/.

26. "In 1987, the NIH Found a Paper Contained Fake Data. It Was Just Retracted," Retraction Watch, October 13, 2022, https://retractionwatch.com/2022/10/13/in-1987-the-nih-found-a-paper-contained-fake-data-it-was-just-retracted/.

27. Retraction Watch, https://retractionwatch.com/.

28. Johnson, Watkinson, and Mabe, *STM Report*, 58.

29. "Global Scientific & Technical Publishing 2019–2023," Research and Markets, https://www.researchandmarkets.com/reports/4912342/global-scientific-and-technical-publishing-2019?utm_source=dynamic&utm_medium=GNOM&utm_code=w67qnp&utm_campaign=1346655+-+Global+%2410B+Scientific+%26+Technical+Publishing+Industry+Report%2c+2019-2023&utm.

30. Software Heritage, https://www.softwareheritage.org/.

31. You can read the declaration here: DORA, "San Francisco Declaration on Research Assessment," https://sfdora.org/read/.

32. D. Moher et al., "The Hong Kong Principles for Assessing Researchers: Fostering Research Integrity," *PLOS Biology* 18, no. 7 (2020): e3000737, https://doi.org/10.1371/journal.pbio.3000737.

Chapter 2

1. P. M. Sutter, G. Lavaux, B. D. Wandelt, and D. H. Weinberg, "A Public Void Catalog from the SDSS DR7 Galaxy Redshift Surveys Based on the Watershed Transform," *Astrophysical Journal* 761, no. 44 (2012).

2. Peter J. Feibelman, *A PhD Is Not Enough!* (New York: Basic Books, 2011).

3. A. Rzhetsky, J. G. Foster, I. T. Foster, and J. A. Evans, "Choosing Experiments to Accelerate Collective Discovery," *PNAS* 112, no. 47 (2015): 14569–74.

4. See "Planck," NASA, https://www.nasa.gov/mission_pages/planck.

5. European Space Agency, "Planck Reveals an Almost Perfect Universe," March 21, 2013, https://www.esa.int/Science_Exploration/Space_Science/Planck/Planck_reveals_an_almost_perfect_Universe.

6. D. B. Kell, "Scientific Discovery as a Combinatorial Optimisation Problem: How Best to Navigate the Landscape of Possible Experiments?," *BioEssays* 34, no. 3 (2012): 236–44.

7. See, for example, National Academies, "Decadal Survey on Astronomy and Astrophysics 2020 (Astro2020)," https://www.nationalacademies.org/our-work/decadal-survey-on-astronomy-and-astrophysics-2020-astro2020.

8. Ethan Siegel and Starts With a Bang, "Why Supersymmetry May Be the Greatest Failed Prediction in Particle Physics History," *Forbes*, February 12, 2019, https://www.forbes.com/sites/startswithabang/2019/02/12/why-supersymmetry-may-be-the-greatest-failed-prediction-in-particle-physics-history/?sh=10849b7869e6.

9. National Cancer Institute, "SEER Cancer Statistics Review (CSR) 1975–2017," April 15, 2020, https://seer.cancer.gov/csr/1975_2017/.

10. Centers for Disease Control and Prevention, "Leading Causes of Death," https://www.cdc.gov/nchs/fastats/leading-causes-of-death.htm.

11. A. Matheson, "Five Steps for Structural Reform in Clinical Cancer Research," *American Journal of Public Health* 100, no. 4 (2010): 596–603.

12. Patrick Collison and Michael Nielsen, "Is Science Stagnant?," *Atlantic*, November 16, 2018, https://www.theatlantic.com/science/archive/2018/11/diminishing-returns-science/575665/; J. Bhattacharya and M. Packalen, "Stagnation and Scientific Incentives," NBER Working Paper 26752, February 2020, http://www.nber.org/papers/w26752.pdf; N. Bloom, C. I. Jones, J. Van Reenen, and M. Webb, "Are Ideas Getting Harder to Find?," *American Economic Review* 110, no. 4 (April 2020): 1104–44.

13. M. Dua, "Scientific Discovery and Its Rationality: Michael Polanyi's Epistemological Exposition," *Foundations of Science* 25, no. 3 (2020): 507–18.

14. S. Arbesman, "Quantifying the Ease of Scientific Discovery," *Scientometrics* 86, no. 2 (2011): 245–50.

15. L. M. A. Bettencourt, D. I. Kaiser, and J. Kaur, "Scientific Discovery and Topological Transitions in Collaboration Networks," *Journal of Informetrics* 3, no. 3 (2009): 210–21.

16. Rzhetsky, Foster, Foster, and Evans, "Choosing Experiments to Accelerate Collective Discovery."

17. Arbesman, "Quantifying the Ease of Scientific Discovery."

18. National Science Foundation, "Universities Report 5.7% Growth in R&D Spending in FY 2019, Reaching $84 Billion," January 13, 2021, https://ncses.nsf.gov/pubs/nsf21313.

19. National Science Board, "NSF Merit Review Reports," https://www.nsf.gov/nsb/publications/pubmeritreview.jsp.

20. Ibid.

21. Mitch Ambrose, "NSF Seeking to Take Risks Despite Flat Budget Proposal," American Institute of Physics, March 20, 2018, https://www.aip.org/fyi/2018/nsf-seeking-take-risks-despite-flat-budget-proposal.

22. American Association for the Advancement of Science (AAAS), "Historical Trends in Federal R&D," https://www.aaas.org/programs/r-d-budget-and-policy/historical-trends-federal-rd.

23. Matt Hourihan and David Parkes, "Energy, Basic Science, Some Space Programs Face Big FY18 Budget Cuts," American Association for the Advancement of Science, May 24, 2017, https://www.aaas.org/news/energy-basic-science-some-space-programs-face-big-fy18-budget-cuts.

24. AAAS, "Historical Trends in Federal R&D."

25. J. Mervis, "Data Check: U.S. Government Share of Basic Research Funding Falls Below 50%," *Science*, March 9, 2017, https://www.science.org/content/article/data-check-us-government-share-basic-research-funding-falls-below-50.

26. NSF, "The State of U.S. Science and Engineering 2020," https://ncses.nsf.gov/pubs/nsb20201/u-s-r-d-performance-and-funding.

27. AAAS, "Historical Trends in Federal R&D"; Larry Smarr, "Trends in Federal Funding of University Research," paper presented at University of California EECS Department Retreat, September 19, 2005, https://www.slideshare.net/Calit2LS/trends-in-federal-funding-of-university-research.

28. National Science Board, "NSF Merit Review Reports."

29. Pew Research Center, "Public and Scientists' Views on Science and Society," January 29, 2015, https://www.pewresearch.org/science/2015/01/29/public-and-scientists-views-on-science-and-society/.

30. National Science Board, "NSF Merit Review Reports."

31. Ibid.

32. Smarr, "Trends in Federal Funding of University Research."

33. National Science Board, "NSF Merit Review Reports."

34. NIH, "Research Project Grants: Competing Applications, Awards, and Success Rates," NIH Data Book, https://report.nih.gov/nihdatabook/report/20.

35. National Science Board, "NSF Merit Review Reports."

36. R. C. Larson, N. Ghaffarzadegan, and M. G. Diaz, "Magnified Effects of Changes in NIH Research Funding Levels," *Service Science* 4, no. 4 (2012): 382–95.

37. National Science Board, "NSF Merit Review Reports."

38. T. von Hippel and C. von Hippel, "To Apply or Not to Apply: A Survey Analysis of Grant Writing Costs and Benefits," *PLOS One* 10 (2015): e0118494, https://doi.org/10.1371/journal.pone.0118494.

39. C. Bloch and M. P. Sørensen, "The Size of Research Funding: Trends and Implications," *Science and Public Policy* 42, no. 1 (2015): 30–43.

40. National Science Board, "NSF Merit Review Reports."

41. S. Copeland, "On Serendipity in Science: Discovery at the Intersection of Chance and Wisdom," *Synthese* 196, no. 6 (2019): 2385–2406.

Chapter 3

1. Noa Maltzman, "Keeping up with Modern Society: Rising Cost of Higher Education," *Medium*, May 8, 2017, https://medium.com/@noamaltzman/keeping-up-with-modern-society-rising-cost-of-higher-education-ce451f052428.

2. P. Stephan, *How Economics Shapes Science* (Cambridge, MA: Harvard University Press, 2012), doi:10.4159/harvard.9780674062757.

3. Ibid.

4. NSF, "Science & Engineering Indicators," https://ncses.nsf.gov/indicators.

5. Ibid.

6. Ibid.

7. Debarghya Das, "The Grad School Admissions Statistics We Never Had," November 17, 2015, https://debarghyadas.com/writes/the-grad-school-statistics-we-never-had/.

8. NSF, "Trends for Graduate Student Enrollment and Postdoctoral Appointments in Science, Engineering, and Health Fields at U.S. Academic Institutions between 2017 and 2019," March 31, 2021, https://ncses.nsf.gov/pubs/nsf21317.

9. Das, "Grad School Admissions Statistics."

10. B. Gemme and Y. Gingras, "Academic Careers for Graduate Students: A Strong Attractor in a Changed Environment," *Higher Education* 63 (2012): 667–83.

11. NSF, "Science & Engineering Indicators."

12. S. Hoenen and C. Kolympiris, "The Value of Insiders as Mentors: Evidence from the Effects of NSF Rotators on Early-Career Scientists," *Review of Economic Statistics* 102, no. 5 (2020): 852–66.

13. G. Laudel and J. Gläser, "From Apprentice to Colleague: The Metamorphosis of Early Career Researchers," *Higher Education* 55 (2008): 387–406.

14. Stephan, *How Economics Shapes Science.*

15. J. P. Walsh and Y. N. Lee, "The Bureaucratization of Science," *Research Policy* 44, no. 8 (2015): 1584–1600.

16. D. K. Simonton, "Scientific Creativity as Constrained Stochastic Behavior: The Integration of Product, Person, and Process Perspectives," *Psychological Bulletin* 129, no. 4 (2003): 475–94, https://doi.org/10.1037/0033-2909.129.4.475.

17. C. Woolston, "PhDs: The Tortuous Truth," *Nature* 575 (2019): 403–6.

18. B. E. Lovitts and C. Nelson, "The Hidden Crisis in Graduate Education: Attrition from Ph.D. Programs," *Academe* 86, no. 6 (2000): 44–50.

19. NSF Science and Engineering Doctorates, "Age at Doctorate Award: What Are the Overall Trends and Characteristics?," https://www.nsf.gov/statistics/2018/nsf18304/report/age-at-doctorate-award-what-are-the-overall-trends-and-characteristics/overall-trends.cfm.

20. NSF Science and Engineering Doctorates, "What Are the Postgraduation Trends?," https://www.nsf.gov/statistics/2018/nsf18304/report/what-are-the-postgraduation-trends/median-salaries.cfm.

21. C. S. Hayter and M. A. Parker, "Factors That Influence the Transition of University Postdocs to Non-academic Scientific Careers: An Exploratory Study," *Research Policy* 48, no. 3 (2019): 556–70.

22. NSF Science and Engineering Doctorates, "Age at Doctorate Award."

23. NSF, "Doctorate Recipients from U.S. Universities: 2019," https://ncses.nsf.gov/pubs/nsf21308.

24. B. L. Benderly, "A Trend toward Transparency for Ph.D. Career Outcomes?," *Science*, March 7, 2018, https://www.science.org/content/article/trend-toward-transparency-phd-career-outcomes.

25. NSF, "Doctorate Recipients from U.S. Universities: 2019," table 42, https://ncses.nsf.gov/pubs/nsf21308/table/42.

26. NSF, "Survey of Doctorate Recipients, 2019," https://ncses.nsf.gov/pubs/nsf21320/table/29.

27. NSF, "Doctorate Recipients," table 42.

28. R. C. Larson, N. Ghaffarzadegan, and Y. Xue, "Too Many PhD Graduates or Too Few Academic Job Openings: The Basic Reproductive Number R0 in Academia," *Systematic Research and Behavioral Science* 31, no. 6 (2014): 745–50.

29. A. M. V. Fournier, A. J. Holford, A. L. Bond, and M. A. Leighton, "Unpaid Work and Access to Science Professions," *PLOS One* 14 (2019), https://doi.org/10.1371/journal.pone.0217032.

30. L. Cruz-Castro and L. Sanz-Menéndez, "Mobility versus Job Stability: Assessing Tenure and Productivity Outcomes," *Research Policy* 39, no. 1 (2010): 27–38.

31. S. Milojevic, F. Radicchi, and J. P. Walsh, "Changing Demographics of Scientific Careers: The Rise of the Temporary Workforce," *PNAS* 115, no. 50 (2018): 12616–23.

32. NSF, "Science & Engineering Indicators"; Maximiliaan Schillebeeckx, Brett Maricque, and Cory Lewis, "The Missing Piece to Changing the University Culture," iAMSTEM HUB, UC Davis, May 28, 2014, https://iamstem.wordpress.com/2014/05/28/the-missing-piece-to-changing-the-university-culture/.

33. Larson, Ghaffarzadegan, and Xue, "Too Many PhD Graduates."

34. NSF, "Science & Engineering Indicators."

35. American Association of University Professors, "Data Snapshot: Contingent Faculty in US Higher Ed," October 11, 2018, https://www.aaup.org/news/data-snapshot-contingent-faculty-us-higher-ed.

36. NSF, "Science & Engineering Indicators."

37. Ibid.

38. Jordan Weissmann, "The Ph.D Bust: America's Awful Market for Young Scientists—in 7 Charts," *Atlantic*, February 20, 2013, https://www.theatlantic.com/business/archive/2013/02/the-phd-bust-americas-awful-market-for-young-scientists-in-7-charts/273339/.

39. NSF, "Science & Engineering Indicators."

40. A. M. Petersen, "Quantifying the Impact of Weak, Strong, and Super Ties in Scientific Careers," *PNAS* 112, no. 34 (2015): E4671–80.

41. NSF, "Science & Engineering Indicators."

42. J. N. Parker, C. Lortie, and S. Allesina, "Characterizing a Scientific Elite: The Social Characteristics of the Most Highly Cited Scientists in Environmental Science and Ecology," *Scientometrics* 85 (2010): 129–43; J. Li, Y. Yin, S. Fortunato, and D. Wang, "Scientific Elite Revisited: Patterns of Productivity, Collaboration, Authorship and Impact," *Journal of the Royal Society Interface* 17, no. 165 (2020), http://doi.org/10.1098/rsif.2020.0135.

43. Stephan, *How Economics Shapes Science*.

44. A. Nowogrodzki, "Most US Professors Are Trained at Same Few Elite Universities," *Nature*, September 21, 2022, https://www.nature.com/articles/d41586-022-02998-w.

45. C. Cañibano, P. D'Este, F. J. Otamendi, and R. Woolley, "Scientific Careers and the Mobility of European Researchers: An Analysis of International Mobility by Career Stage," *Higher Education* 80 (2020): 1175–93.

46. Isaiah Hankel, "5 Career Killing Mistakes PhDs Make (#4 Is Very Common)," LinkedIn, July 30, 2018, https://www.linkedin.com/pulse/5-career-killing-mistakes-phds-make-4-very-common-hankel-ph-d-/.

47. Jonathon Wosen, "Exodus of Young Life Scientists Is Shaking Up Academia," *STAT*, November 10, 2020, https://www.statnews.com/2022/11/10/tipping-point-is-coming-unprecedented-exodus-of-young-life-scientists-shaking-up-academia/.

48. J. Gould, "Planning a Postdoc before Moving to Industry? Think Again," *Nature*, December 3, 2020, https://www.nature.com/articles/d41586-020-03109-3.

49. DORA, "San Francisco Declaration on Research Assessment," https://sfdora.org/read/.

50. Rutgers School of Graduate Studies, "Biomedical Career Development," https://grad.rutgers.edu/professional-development/biomedical-career-development.

51. NYU Langone Health, "Job Search & Career Transition for Graduate Students & Postdoctoral Fellows," https://med.nyu.edu/research/postdoctoral-affairs/professional-development/job-search-career-transition.

52. National Institutes of Health, Office of Intramural Training and Education, "Graduate Students," https://www.training.nih.gov/career_services/graduate_students.

53. Scitable, "Graduate School, Further Training," https://www.nature.com/scitable/topicpage/graduate-school-and-further-training-14018592/.

54. Erdős Institute, https://www.erdosinstitute.org/.

55. Berkeley Career Center, "Career Fairs," https://career.berkeley.edu/Fairs/Fairs; University of Colorado Boulder, Career Services, "How Grad Students Can Prepare for Career Fairs," September 28, 2022, https://www.colorado.edu/career/2022/09/28/how-grad-students-can-prepare-career-fairs.

56. Alan Leshner and Layne Scherer, *Graduate STEM Education for the 21st Century* (Washington, DC: National Academies Press, 2018), doi:10.17226/25038.

57. Elsevier, "University-Industry Collaboration: A Closer Look for Research Leaders," January 27, 2021, https://www.elsevier.com/research-intelligence/university-industry-collaboration; Univesrity of Minnesota, Office of the Vice President for Research, "Industry Partnerships for Researchers," https://research.umn.edu/industry-partnership/researchers; University of Maryland, Behavioral & Social Sciences College, "University-Industry Partnerships in the Social Sciences: Helping Organizations Achieve Impact," https://bsos.umd.edu/academics-research/university-industry.

Chapter 4

1. E. C. M. Parsons, "'Advocacy' and 'Activism' Are Not Dirty Words—How Activists Can Better Help Conservation Scientists," *Frontiers in Marine Science* 3, no. 229 (2016).

2. David C. Lindberg and Ronald L. Numbers, "Beyond War and Peace: A Reappraisal of the Encounter between Christianity and Science," *Perspectives on Science and Christian Faith* 39, no. 3 (1987): 140–49, https://www.asa3.org/ASA/PSCF/1987/PSCF9-87Lindberg.html.

3. Pew Research Center, "Public Praises Science; Scientists Fault Public, Media—Section 4: Scientists, Politics and Religion," July 9, 2009, https://www.pewresearch.org/politics/2009/07/09/section-4-scientists-politics-and-religion/.

4. Cary Funk, Meg Hefferon, Brian Kennedy, and Courtney Johnson, "Trust and Mistrust in Americans' Views of Scientific Experts," Pew Research Center, August 2, 2019, https://www.pewresearch.org/science/2019/08/02/trust-and-mistrust-in-americans-views-of-scientific-experts/.

5. Pew Research Center, "The New Food Fights: U.S. Public Divides over Food Science—Section 3: Public Opinion about Genetically Modified Foods and Trust in Scientists," December 1, 2016, https://www.pewresearch.org/science/2016/12/01/public-opinion-about-genetically-modified-foods-and-trust-in-scientists-connected-with-these-foods/.

6. Joshua Rapp Learn, "Canadian Scientists Explain Exactly How Their Government Silenced Science," *Smithsonian Magazine*, January 30, 2017, https://www.smithsonianmag.com/science-nature/canadian-scientists-open-about-how-their-government-silenced-science-180961942/.

7. Pew Research Center, "Politics and Science: What Americans Think," July 1, 2015, https://www.pewresearch.org/science/2015/07/01/americans-politics-and-science-issues/.

8. *Smoking and Health: Report of the Advisory Committee to the Surgeon General of the Public Health Service* (Washington, DC: United States Department of Health, Education, and Welfare, 1964), https://profiles.nlm.nih.gov/spotlight/nn/catalog/nlm:nlmuid-101584932X202-doc.

9. Stanton A. Glantz, John Slade, Lisa A. Bero, Peter Hanauer, and Deborah E Barnes, eds., *The Cigarette Papers* (Berkeley: University of California Press, 1998).

10. InTeGrate, "542 Million Years of Sea Level Change: Exxon's Sea Level Reconstruction," https://serc.carleton.edu/integrate/teaching_materials/coastlines/student_materials/890.

11. John Vidal, "Revealed: How Oil Giant Influenced Bush," *Guardian*, June 8, 2005, https://www.theguardian.com/news/2005/jun/08/usnews.climatechange.

12. B. Sosa, E. Fontans-Álvarez, D. Romero, A. da Fonseca, and M. Achkar, "Analysis of Scientific Production on Glyphosate: An Example of Politicization of Science," *Science of the Total Environment* 681 (2019): 541–50.

13. MyPlate, https://www.choosemyplate.gov/.

14. N. Teicholz, "The Scientific Report Guiding the US Dietary Guidelines: Is It Scientific?," *BMJ* 351 (2015).

15. Sarah Kaplan, "Are Scientists Going to March on Washington?," *Washington Post*, January 25, 2017, https://www.washingtonpost.com/news/speaking-of-science/wp/2017/01/24/are-scientists-going-to-march-on-washington/.

16. March for Science, https://marchforscience.org/.

17. March for Science, "About Us," https://web.archive.org/web/20170318183426/https://www.marchforscience.com/mission-and-vision.

18. March for Science, "The Time Has Come," https://web.archive.org/web/20201126172049/https://marchforscience.org/the-time-has-come/.

19. Democratic National Committee, "Combating the Climate Crisis and Pursuing Environmental Justice," https://democrats.org/where-we-stand/party-platform/combating-the-climate-crisis-and-pursuing-environmental-justice/.

20. R. A. Pielke, *The Honest Broker: Making Sense of Science in Policy and Politics* (Cambridge: Cambridge University Press, 2007).

21. A. Staudt, N. Huddleston, and I. Kraucunas, "Understanding and Responding to Climate Change: Highlights of National Academies Reports," National Academies of Science, 2008, http://dels-old.nas.edu/dels/rpt_briefs/climate_change_2008_final.pdf.

22. Holly Yan, "Want to Prevent Another Shutdown, Save 33,000 Lives and Protect Yourself? Wear a Face Mask, Doctors Say," CNN, June 29, 2020, https://www.cnn.com/2020/06/25/health/face-mask-guidance-covid-19/index.html.

23. H. Schmid-Petri, "Politicization of Science: How Climate Change Skeptics Use Experts and Scientific Evidence in Their Online Communication," *Climatic Change* 145, no. 3 (2017): 523–37.

24. M. Tesler, "Elite Domination of Public Doubts about Climate Change (Not Evolution)," *Political Communication* 35, no. 2 (2018): 306–26.

25. COVID-19 Projections Using Machine Learning, "Model Comparison with IHME," https://covid19-projections.com/model-comparison-ihme.

26. COVID-19 Projections Using Machine Learning, "About COVID19-Projections.com," https://covid19-projections.com/about/#comparison-of-october-us-projections.

27. Pew Research Center, "Public Praises Science."

28. Pew Research Center, "Politics and Science."

29. S. van der Linden, A. Leiserowitz, S. Rosenthal, and E. Maibach, "Inoculating the Public against Misinformation about Climate Change," *Global Challenges* 1, no. 2 (2017): 1600008.

30. R. A. Pielke, "When Scientists Politicize Science: Making Sense of Controversy over *The Skeptical Environmentalist*," *Environmental Science & Policy* 7 (2004): 405–17.

31. Cary Funk, Meg Hefferon, Brian Kennedy, and Courtney Johnson, "Trust and Mistrust in Americans' Views of Scientific Experts—Section 1: Partisanship Influences Views on the Role, Value of Scientific Experts," Pew Research Center, August 2, 2019, https://www.pewresearch.org/science/2019/08/02/partisanship-influences-views-on-the-role-and-value-of-scientific-experts-in-policy-debates/.

32. "Trust in Science Is Becoming More Polarized, Survey Finds," *University of Chicago News*, January 28, 2022, https://news.uchicago.edu/story/trust-science-becoming-more-polarized-survey-finds.

33. Brian Kennedy and Cary Funk, "Partisans Differ over Scientists' Role and Value in Policy Debates," Pew Research Center, August 9, 2019, https://www.pewresearch.org/fact-tank/2019/08/09/democrats-and-republicans-role-scientists-policy-debates/.

34. Cary Funk, Meg Hefferon, Brian Kennedy, and Courtney Johnson, "Trust and Mistrust in Americans' Views of Scientific Experts—Section 2: Americans Often Trust Practitioners More Than Researchers," Pew Research Center, August 2, 2019, https://www.pewresearch.org/science/2019/08/02/americans-often-trust-practitioners-more-than-researchers-but-are-skeptical-about-scientific-integrity/.

35. J. E. Kotcher, T. A. Myers, E. K. Vraga, N. Stenhouse, and E. W. Maibach, "Does Engagement in Advocacy Hurt the Credibility of Scientists? Results from a Randomized National Survey Experiment," *Environmental Communication* 11, no. 3 (2017): 415–29.

36. N. A. Rose and E. C. M. Parsons, "'Back Off, Man, I'm a Scientist!': When Marine Conservation Science Meets Policy," *Ocean & Coastal Management* 115 (2015): 71–76.

37. Ivan Oransky and Adam Marcus, "Should Scientists Engage in Activism?," *Conversation*, February 6, 2017, https://theconversation.com/should-scientists-engage-in-activism-72234; Alyssa Shearer, Ingrid Joylyn Paredes, Tiara Ahmad, and Christopher Jackson, "Yes, Science Is Political," *Scientific American*, October 8, 2020, https://www.scientificamerican.com/article/yes-science-is-political/; Sarah Kaplan, "Scientists Prepare to Fight for Their Work During 'the Trumpocene,'" *Washington Post*, December 15, 2016, https://www.washingtonpost.com/news/speaking-of-science/wp/2016/12/15/researchers-reckon-with-the-trumpocene-at-the-worlds-largest-earth-science-meeting/; Fernando Tormos-Aponte, Scott Frickel, and John Parker, "Scientists Are Becoming More Politically Engaged," *Scientific American*, November 25, 2020, https://www.scientificamerican.com/article/scientists-are-becoming-more-politically-engaged/; Becca Muir, "Should Scientists Be Activists?," *Chemistry World*, March 18, 2020, https://www.chemistryworld.com/careers/should-scientists-be-activists/4011293.article; Matthew C. Nisbet, "Do Scientists Have a Special Responsibility to Engage in Political Advocacy?," *Big Think*, September 13, 2010, https://bigthink.com/guest-thinkers/do-scientists-have-a-special-responsibility-to-engage-in-political-advocacy/.

38. D. Sedlak, "Crossing the Imaginary Line," *Environmental Science & Technology* 50, no. 18 (2016): 9803–4.

Chapter 5

1. R. Gasparatou, "Scientism and Scientific Thinking," *Science & Education* 26 (2017): 799–812; M. Stenmark, "What Is Scientism?," *Religious Studies* 33 (1997): 15–32; R. Peels, "Ten Reasons to Embrace Scientism," *Studies in History and Philosophy of Science Part A* 63 (2017): 11–21.

2. H. C. Shulman, G. N. Dixon, O. M. Bullock, and D. Colón Amill, "The Effects of Jargon on Processing Fluency, Self-Perceptions, and Scientific Engagement," *Journal of Language and Social Psychology* 39, nos. 5–6 (2020): 579–97.

3. O. M. Bullock, D. Colón Amill, H. C. Shulman, and G. N. Dixon, "Jargon as a Barrier to Effective Science Communication: Evidence from Metacognition," *Public Understanding of Science* 28, no. 7 (2019): 845–53.

4. Massimo Pigliucci, "Neil deGrasse Tyson and the Value of Philosophy," *Scientia Salon*, May 12, 2014, https://scientiasalon.wordpress.com/2014/05/12/neil-degrasse-tyson-and-the-value-of-philosophy/.

5. Matt Warman, "Stephen Hawking Tells Google 'Philosophy Is Dead,'" *Telegraph*, May 17, 2011, https://www.telegraph.co.uk/technology/google/8520033/Stephen-Hawking-tells-Google-philosophy-is-dead.html.

6. Olivia Goldhill, "Bill Nye: Why Are So Many Smart People Such Idiots about Philosophy?," *Quartz*, March 5, 2016, https://qz.com/627989/why-are-so-many-smart-people-such-idiots-about-philosophy/.

7. L. Laplane et al., "Why Science Needs Philosophy," *PNAS* 116, no. 10 (2019): 3948–52, https://doi.org/10.1073/pnas.1900357116.

8. Olivia Goldhill, "Bill Nye on Philosophy: The Science Guy Says He Has Changed His Mind," *Quartz*, April 15, 2017, https://qz.com/960303/bill-nye-on-philosophy-the-science-guy-says-he-has-changed-his-mind/.

9. A. Gare, "Natural Philosophy and the Sciences: Challenging Science's Tunnel Vision," *Philosophies* 3, no. 4 (2018).

10. Jackson Lears, "Get Happy!!," *Nation*, November 25, 2013, https://www.thenation.com/article/archive/get-happy-2/#; Thomas Nagel, "The Facts Fetish," *New Republic*, October 20, 2010, https://newrepublic.com/article/78546/the-facts-fetish-morality-science; John Gray, "The Knowns and the Unknowns," *New Republic*, April 20, 2012, https://newrepublic.com/article/102760/righteous-mind-haidt-morality-politics-scientism; Robert C. Bannister, *Sociology and Scientism: The American Quest for Objectivity, 1880–1940* (Chapel Hill: University of North Carolina Press, 1991); Leon Wieseltier, "Crimes against Humanities," *New Republic*, September 3, 2013, https://newrepublic.com/article/114548/leon-wieseltier-responds-steven-pinkers-scientism.

11. B. E. Kaufman, "The Real Problem: The Deadly Combination of Psychologisation, Scientism, and Normative Promotionalism Takes Strategic Human Resource Management down a 30-Year Dead End," *Human Resource Management Journal* 30, no. 1 (2020): 49–72.

12. A. B. Bruce, "Frankenfish or Fish to Feed the World? Scientism and Biotechnology Regulatory Policy," *Rural Sociology* 82, no. 4 (2017): 628–63.

13. C. R. Mayes and D. B. Thompson, "What Should We Eat? Biopolitics, Ethics, and Nutritional Scientism," *Journal of Bioethical Inquiry* 12, no. 4 (2015): 587–99.

14. L. Reid, "Scientism in Medical Education and the Improvement of Medical Care: Opioids, Competencies, and Social Accountability," *Health Care Analysis* 26, no. 2 (2018): 155–70.

15. G. Blue, "Scientism: A Problem at the Heart of Formal Public Engagement with Climate Change," *ACME* 17, no. 2 (2018): 544–60.

16. C. Ohlers, "The 'Conflict Thesis' of Science and Religion: A Nexus of Philosophy of Science, Metaphysics, and Philosophy of Religion," *Journal of Biblical and Theological Studies* 2, no. 2 (2017): 208–33; J. C. Ungureanu, "'Your God Is Too Small': Retracing the Origins of Conflict between Science and Religion," *Theology and Science* 20, no. 1 (2022): 23–45.

17. Ki Mae Heussner, "Stephen Hawking on Religion: 'Science Will Win,'" ABC News, June 4, 2010, https://abcnews.go.com/WN/Technology/stephen-hawking-religion-science-win/story?id=10830164.

18. Carl Sagan, *The Demon-Haunted World: Science as a Candle in the Dark* (New York: Random House, 1995).

19. Stephen Jay Gould, *Rocks of Ages: Science and Religion in the Fullness of Life* (New York: Penguin, 1999).

20. E. H. Ecklund and J. Z. Park, "Conflict between Religion and Science among Academic Scientists?," *Journal for the Scientific Study of Religion* 48, no. 2 (2009): 276–92.

21. J. Astley and L. J. Francis, "Promoting Positive Attitudes towards Science and Religion among Sixth-Form Pupils: Dealing with Scientism and Creationism," *British Journal of Religious Education* 32 (2010): 189–200.

22. Gary B. Ferngren, ed., *Science and Religion: A Historical Introduction*, 2nd ed. (Baltimore, MD: Johns Hopkins University Press, 2017).

23. "The Galileo Myths," Scientus.org, https://www.scientus.org/Galileo-Myths.html.

24. Richard W. Pogge, "The Folly of Giordano Bruno," http://www.astronomy.ohio-state.edu/~pogge/Essays/Bruno.html.

25. Seb Falk, *The Light Ages: The Surprising Story of Medieval Science* (New York: W. W. Norton, 2020), 391.

26. J. H. Evans, "Epistemological and Moral Conflict between Religion and Science," *Journal for the Scientific Study of Religion* 50, no. 4 (2011): 707–27.

27. Pew Research Center, "Religious and Science," October 22, 2015, https://www.pewresearch.org/science/2015/10/22/science-and-religion/.

28. Ibid.

29. Rick Mullin, "Behind the Scenes at the STEM-Humanities Culture War," *C&EN* 97 (2019): 26–29; "Science vs. Humanities: Educating Citizens of the Future," *Elesapiens' Blog*, September 11, 2014, https://www.elesapiens.com/blog/science-vs-humanities-educating-citiziens-of-the-future/.

30. P. Jay, "The Humanities Crisis Then and Now," in *The Humanities "Crisis" and the Future of Literary Studies*, 7–31 (New York: Palgrave Macmillan, 2014).

31. Benjamin Schmidt, "The Humanities Are in Crisis," *Atlantic*, August 23, 2018, https://www.theatlantic.com/ideas/archive/2018/08/the-humanities-face-a-crisisof-confidence/567565/.

32. A. Ruben, "Scientists Should Defend, Not Defund, the Humanities," *Science*, September 23, 2015, https://www.science.org/content/article/scientists-should-defend-not-defund-humanities; Lior Shamir, "A Case against the STEM Rush," *Inside Higher Ed*, February 2, 2020, https://www.insidehighered.com/views/2020/02/03/computer-scientist-urges-more-support-humanities-opinion; Jenann Ismael, "Why Science Will Never Replace the Humanities,"

Philosophy Talk, December 12, 2015, https://www.philosophytalk.org/blog/why-science-will-never-replace-humanities; Maria Konnikova, "Humanities Aren't a Science. Stop Treating Them Like One," *Scientific American*, August 10, 2012, https://blogs.scientificamerican.com/literally-psyched/humanities-arent-a-science-stop-treating-them-like-one/.

Chapter 6

1. Scott Williams, "Edward Alexander Bouchet," Physicists of the African Diaspora, http://www.math.buffalo.edu/mad/physics/bouchet_edward_alexander.html.
2. NSF, "Doctorate Recipients from U.S. Universities: 2021," https://ncses.nsf.gov/pubs/nsf23300/report/u-s-doctorate-awards#race-and-ethnicity.
3. NSF, "Who Earns a U.S. Doctorate?," Science and Engineering Doctorates, https://www.nsf.gov/statistics/2018/nsf18304/report/who-earns-a-us-doctorate/race-and-ethnicity.cfm.
4. National Society of Black Physicists, "Utilization of African-American Physicists in the Science and Engineering Workforce," in *Pan-Organizational Summit on the US Science and Engineering Workforce*, ed. Marye Anne Fox (Washington, DC: National Academies Press, 2003), https://www.ncbi.nlm.nih.gov/books/NBK36368/.
5. NSF, "Doctorate Recipients from U.S. Universities: 2021."
6. Adam Harris, "Why Are There So Few Black Ph.D.s?," *Atlantic*, April 19, 2019, https://www.theatlantic.com/education/archive/2019/04/lack-of-black-doctoral-students/587413/.
7. NSF, "Doctorate Recipients from U.S. Universities: 2021."
8. NSF, "Women, Minorities, and Persons with Disabilities in Science and Engineering: 2021," https://ncses.nsf.gov/pubs/nsf21321/.
9. American Institute of Physics, "Data on Underrepresented Groups among Undergraduates," https://www.aip.org/statistics/undergraduate/minorities.
10. NSF, "Women, Minorities, and Persons with Disabilities in Science and Engineering: 2021."
11. Ibid.
12. U.S. Census Bureau, "QuickFacts: United States," https://www.census.gov/quickfacts/fact/table/US/IPE120221.
13. Megan Gannon, "Race Is a Social Construct, Scientists Argue," *Scientific American*, February 5, 2019, https://www.scientificamerican.com/article/race-is-a-social-construct-scientists-argue/.

14. Christopher Jencks and Meredith Phillips, "The Black-White Test Score Gap: Why It Persists and What Can Be Done," Brookings, March 1, 1998, https://www.brookings.edu/articles/the-black-white-test-score-gap-why-it-persists-and-what-can-be-done/.

15. American Association of Biological Anthropologists, "AABA Statement on Race & Racism," https://physanth.org/about/position-statements/aapa-statement-race-and-racism-2019/.

16. NSF, "Who Earns a U.S. Doctorate?"

17. J. J. Gottlieb, "STEM Career Aspirations in Black, Hispanic, and White Ninth-Grade Students," *Journal of Research in Science Teaching* 55, no. 2 (2018): 1365–92.

18. Y. Copur-Gencturk, J. R. Cimpian, S. T. Lubienski, and I. Thacker, "Teachers' Bias against the Mathematical Ability of Female, Black, and Hispanic Students," *Educational Researcher* 49, no. 1 (2020): 30–43.

19. C. M. Ganley, C. E. George, J. R. Cimpian, and M. B. Makowski, "Gender Equity in College Majors: Looking Beyond the STEM/Non-STEM Dichotomy for Answers Regarding Female Participation," *American Educational Research Journal* 55, no. 3 (2018): 453–87.

20. S. L. Kuchynka et al., "Hostile and Benevolent Sexism and College Women's STEM Outcomes," *Psychology of Women Quarterly* 42, no. 1 (2018): 72–87.

21. National Academies Press, "Evidence-Based Interventions for Addressing the Underrepresentation of Women in Science, Engineering, Mathematics, and Medicine," 2020, doi:10.17226/25786.; S.-J. Leslie, A. Cimpian, M. Meyer, and E. Freeland, "Expectations of Brilliance Underlie Gender Distributions across Academic Disciplines," *Science* 347 (2015): 262–65.

22. A. Garr-Schultz and W. L. Gardner, "Strategic Self-Presentation of Women in STEM," *Social Sciences* 7, no. 2 (2018): article 20.

23. National Academies Press, *The Science of Effective Mentorship in STEMM* (Washington, DC: National Academies Press, 2019), doi:10.17226/25568.

24. National Academies Press, *An American Crisis: The Growing Absence of Black Men in Medicine and Science* (Washington, DC: National Academies Press, 2018), doi:10.17226/25130.

25. M. C. Saffie-Robertson, "It's Not You, It's Me: An Exploration of Mentoring Experiences for Women in STEM," *Sex Roles* 83, no. 4 (2020): 533–79.

26. National Academies Press, "Evidence-Based Interventions"; Leslie, Cimpian, Meyer, and Freeland, "Expectations of Brilliance Underlie Gender Distributions."

27. National Academies Press, *Sexual Harassment of Women: Climate, Culture, and Consequences in Academic Sciences, Engineering, and Medicine* (Washington, DC: National Academies Press, 2018), doi:10.17226/24994.

28. K. Brue, "Work-Life Balance for Women in STEM Leadership," *Journal of Leadership Education* 18 (2019): 32–52.

29. Felix Richter, "The Nobel Prize Gender Gap," Statista, October 13, 2020, https://www.statista.com/chart/2805/nobel-prize-winners-by-gender/.

30. R. D. Robnett and S. E. Thoman, "STEM Success Expectancies and Achievement among Women in STEM Majors," *Journal of Applied Developmental Psychology* 52 (2017): 91–100; B. Bloodhart, M. M. Balgopal, A. M. A. Casper, L. B. Sample McMeeking, and E. V. Fischer, "Outperforming yet Undervalued: Undergraduate Women in STEM," *PLOS One* 15, no. 6 (2020): e0234685, https://doi.org/10.1371/journal.pone.0234685.

31. L. McCullough, "Proportions of Women in STEM Leadership in the Academy in the USA," *Educational Sciences* 10 , no. 1 (2020): 1–13.

32. National Academies Press, *Science of Effective Mentorship in STEMM*.

33. P. Akos and J. Kretchmar, "Gender and Ethnic Bias in Letters of Recommendation: Considerations for School Counselors," *Professional School Counseling* 20, no. 1 (2016): 102–13.

34. T. A. Hoppe et al., "Topic Choice Contributes to the Lower Rate of NIH Awards to African-American/Black Scientists," *Science Advances* 5, no. 10 (2019).

35. National Academies Press, *Sexual Harassment of Women*.

36. R. van Veelen, B. Derks, and M. D. Endedijk, "Double Trouble: How Being Outnumbered and Negatively Stereotyped Threatens Career Outcomes of Women in STEM," *Frontiers in Psychology* 10 (2019), https://doi.org/10.3389/fpsyg.2019.00150.

37. F. Danbold and Y. J. Huo, "Men's Defense of Their Prototypicality Undermines the Success of Women in STEM Initiatives," *Journal of Experimental Social Psychology* 72 (2017): 57–66.

38. National Academies Press, *Sexual Harassment of Women*.

39. "Geoffrey Marcy's Berkeley Astronomy Colleagues Call for His Dismissal," *New York Times*, October 14, 2015, https://www.nytimes.com/2015/10/14/science/geoffrey-marcy-berkeley-astronomy-faculty-letter.html.

40. Akos and Kretchmar, "Gender and Ethnic Bias in Letters of Recommendation."

41. Scott Jaschik, "Do Recommendation Letters Insert Bias into College Admissions Decisions?," *Inside Higher Ed*, October 21, 2018, https://www.insidehighered.com/admissions/article/2018/10/22/do-recommendation-letters-insert-bias-college-admissions-decisions; Michelle Irwin, "Letters of Recommendation Reaffirm Entrenched Systems of Bias and Exclusion," *Inside Higher Ed*, April 9, 2019, https://www.insidehighered.com/advice/2019/04/10/letters-recommendation-reaffirm-entrenched-systems-bias-and-exclusion-opinion;

T. Thornhill, "We Want Black Students, Just Not You: How White Admissions Counselors Screen Black Prospective Students," *Sociology of Race and Ethnicity* 5 (2019): 456–70.

42. T. Leslie, "'I Decided': Listening to the Voices of Academically Resilient Black and Hispanic Students in College STEM Programs," *Dissertation Abstracts International Section A: Humanities and Social Sciences* 77 (2017).

43. T. R. Sass, "Understanding the STEM Pipeline," National Center for Analysis of Longitudinal Data in Education Research, Calder Working Paper 125, January 2015, https://files.eric.ed.gov/fulltext/ED560681.pdf.

44. G. Saw, C. N. Chang, and H. Y. Chan, "Cross-Sectional and Longitudinal Disparities in STEM Career Aspirations at the Intersection of Gender, Race/Ethnicity, and Socioeconomic Status," *Educational Researcher* 47, no. 8 (2018): 525–31.

45. C. M. Harris, "Quitting Science: Factors That Influence Exit from the STEM Workforce," *Journal of Women and Minorities in Science and Engineering* 25, no. 2 (2019): 93–118.

46. National Academies Press, *Science of Effective Mentorship in STEMM.*

47. B. M. Dewsbury, C. Taylor, A. Reid, and C. Viamonte, "Career Choice among First-Generation, Minority STEM College Students," *Journal of Microbiology & Biology Education* 20, no. 3 (2019).

48. S. Nealy and M. K. Orgill, "Postsecondary Underrepresented Minority STEM Students' Perceptions of Their Science Identity," *Journal of Negro Education* 88, no. 3 (2019): 249–68.

49. A. J. Fisher et al., "Structure and Belonging: Pathways to Success for Underrepresented Minority and Women PhD Students in STEM Fields," *PLOS One* 14, no. 1 (2019): e0209279, https://doi.org/10.1371/journal.pone.0209279.

50. National Academies Press, *Minority Serving Institutions: America's Underutilized Resource for Strengthening the STEM Workforce* (Washington, DC: National Academies Press, 2019), doi:10.17226/25257.

51. S. Williams-Watson, "Persistence among Minority STEM Majors: A Phenomenological Study," *ProQuest Dissertations and Theses* (2017).

52. A. Meador, "Examining Recruitment and Retention Factors for Minority STEM Majors through a Stereotype Threat Lens," *School Science & Mathematics* 118, nos. 1–2 (2018): 61–69.

53. National Academies Press, "Evidence-Based Interventions"; Leslie, Cimpian, Meyer, and Freeland, "Expectations of Brilliance Underlie Gender Distributions."

54. D. D. Rupert, A. C. Nowlan, O. H. Tam, and M. G. Hammell, "Ten Simple Rules for Running a Successful Women-in-STEM Organization on an Academic Campus," *PLOS Computational Biology* 16, no. 5 (2020): e1007754, https://doi.org/10.1371/journal.pcbi.1007754.

55. National Academies Press, "Evidence-Based Interventions"; Leslie, Cimpian, Meyer, and Freeland, "Expectations of Brilliance Underlie Gender Distributions."

56. N. Thomas, J. Bystydzienski, and A. Desai, "Changing Institutional Culture through Peer Mentoring of Women STEM Faculty," *Innovative Higher Education* 40, no. 2 (2015): 143–57.

57. L. R. Ramsey, D. E. Betz, and D. Sekaquaptewa, "The Effects of an Academic Environment Intervention on Science Identification among Women in STEM," *Social Psychology of Education* 16, no. 3 (2013): 377–97.

58. A. M. Ryan, D. D. King, F. Elizondo, and P. Wadlington, "Social Identity Management Strategies of Women in STEM Fields," *Journal of Occupational & Organizational Psychology* 93, no. 2 (2020): 245–72.

59. A. Yadav et al., "The Forgotten Scholar: Underrepresented Minority Postdoc Experiences in STEM Fields," *Educational Studies* 56, no. 2 (2020): 160–85.

60. M. Estrada et al., "Improving Underrepresented Minority Student Persistence in STEM," *CBE—Life Sciences Education* 15, no. 3 (2016).

61. E. A. Canning, K. Muenks, D. J. Green, and M. C. Murphy, "STEM Faculty Who Believe Ability Is Fixed Have Larger Racial Achievement Gaps and Inspire Less Student Motivation in Their Classes," *Science Advances* 5, no. 2 (2019).

62. D. C. Martin, "It's Not My Party: A Critical Analysis of Women and Minority Opposition towards STEM," *Critical Questions in Education* 7, no. 2 (2016): 96–115.

Chapter 7

1. Julia Belluz, "Why So Many of the Health Articles You Read Are Junk," *Vox*, December 10, 2014, https://www.vox.com/2014/12/10/7372921/health-journalism-science.

2. Nick Tate, "Bacon, Ham, Sausage as Dangerous as Smoking: Report," Newsmax.com, October 23, 2015, https://www.newsmax.com/Health/Diet-And-Fitness/bacon-sausage-processed-meat/2015/10/23/id/698726/.

3. Michael Morrow, "Bearded Men Have Poop on Their Faces," *New York Post*, May 4, 2015, https://nypost.com/2015/05/04/science-proves-that-beards-contain-fecal-matter/.

4. Megan Friedman, "Your Nail Polish Might Be Making You Gain Weight," *Elle*, October 20, 2015, https://www.elle.com/beauty/makeup-skin-care/news/a31283/tphp-nail-polish-study/.

5. Doyle Rice, "Climate Change Could Be Causing You to Lose Sleep, Study Finds," *USA Today*, May 26, 2017, https://www.usatoday.com/story/weather/2017/05/26/climate-change-causes-lost-sleep/102148614/.

6. Avi Selk, "Harvard's Top Astronomer Says an Alien Ship May Be among Us—and He Doesn't Care What His Colleagues Think," *Washington Post*, February 4, 2019, https://www.washingtonpost.com/lifestyle/style/harvards-top-astronomer-says-an-alien-ship-may-be-among-us—and-he-doesnt-care-what-his-colleagues-think/2019/02/04/a5d70bb0-24d5-11e9-90cd-dedb0c92dc17_story.html.

7. Matthew Diebel, "Full-Fat Dairy Milk, Cheese and Yogurt Isn't Bad for You, Study Finds," *USA Today*, May 9, 2017, https://www.usatoday.com/story/news/health/2017/05/09/full-fat-dairy-stuff——cheese-yogurt-and-so——isnt-bad-you-study-finds/101461372/.

8. A. Moore, "Bad Science in the Headlines. Who Takes Responsibility When Science Is Distorted in the Mass Media?," *EMBO Reports* 7, no. 12 (2006): 1193–96.

9. L. F. Santos, "The Role of Critical Thinking in Science Education," *Journal of Education and Practice* 8, no. 20 (2017): 159–73; R. A. Larimore, "Preschool Science Education: A Vision for the Future," *Early Childhood Education Journal* 48 (2020): 703–14, https://doi.org/10.1007/s10643-020-01033-9; D. B. Larkin, "Attending to the Public Understanding of Science Education: A Response to Furtak and Penuel," *Science Education* 103, no. 5 (2019): 1294–1300; M. Lindholm, "Promoting Curiosity? Possibilities and Pitfalls in Science Education," *Science & Education* 27, nos. 9–10 (2018): 987–1002.

10. S. Laursen, C. Liston, H. Thiry, and J. Graf, "What Good Is a Scientist in the Classroom? Participant Outcomes and Program Design Features for a Short-Duration Science Outreach Intervention in K–12 Classrooms," *CBE Life Sciences Education* 6, no. 1 (2007): 49–64.

11. E. H. Ecklund, S. A. James, and A. E. Lincoln, "How Academic Biologists and Physicists View Science Outreach," *PLOS One* 7, no. 5 (2012): e36240, https://doi.org/10.1371/journal.pone.0036240; J. Varner, "Scientific Outreach: Toward Effective Public Engagement with Biological Science," *BioScience* 64, no. 4 (2014): 333–40, https://doi.org/10.1093/biosci/biu021.

12. C. McClain and L. Neeley, "A Critical Evaluation of Science Outreach via Social Media: Its Role and Impact on Scientists," *F1000Research* 3 (2015), https://doi.org/10.12688/f1000research.5918.1.

13. C. R. McClain, "Practices and Promises of Facebook for Science Outreach: Becoming a 'Nerd of Trust,'" *PLOS Biology* 15, no. 6 (2017): e2002020, https://doi.org/10.1371/journal.pbio.2002020.

14. S. Martinez-Conde, "Has Contemporary Academia Outgrown the Carl Sagan Effect?," *Journal of Neuroscience* 36, no. 7 (2016): 2077–82, https://doi.org/10.1523/JNEUROSCI.0086-16.2016.

15. Ibid.

16. NSF, "Broader Impacts," https://www.nsf.gov/od/oia/special/broaderimpacts/.

17. N. M. Nadkarni and A. E. Stasch, "How Broad Are Our Broader Impacts? An Analysis of the National Science Foundation's Ecosystem Studies Program and the Broader Impacts Requirement," *Frontiers in Ecology and the Environment* 11, no. 1 (2013): 13–19.

18. M. R. Roberts, "Realizing Societal Benefit from Academic Research: Analysis of the National Science Foundation's Broader Impacts Criterion," *Social Epistemology* 23, nos. 3–4 (2009): 199–219.

19. Jacob Carter, "The American Public Still Trusts Scientists, Says a New Pew Survey," *Scientific American*, September 29, 2020, https://www.scientificamerican.com/article/the-american-public-still-trusts-scientists-says-a-new-pew-survey/.

20. L. Smith-Doerr, "Hidden Injustice and Anti-science," *Engaging Science, Technology, and Society* 6 (2020): 94–101.

21. S. O. Hansson, "Dealing with Climate Science Denialism: Experiences from Confrontations with Other Forms of Pseudoscience," *Climate Policy* 18, no. 9 (2018): 1094–1102.

22. Ben Goldacre, *Bad Science: Quacks, Hacks, and Big Pharma Flacks* (New York: Farrar, Straus & Giroux, 2011).

23. B. T. Rutjens, R. M. Sutton, and R. van der Lee, "Not All Skepticism Is Equal: Exploring the Ideological Antecedents of Science Acceptance and Rejection," *Personality & Social Psychology Bulletin* 44, no. 3 (2018): 384–405.

24. D. Scales, S. Gorman, and J. Gorman, "Resist Pseudoscience with Respect, Not Ridicule," *Nature*, June 2, 2020, https://www.nature.com/articles/d41586-020-01626-9.

25. B. A. Zaboski and D. J. Therriault, "Faking Science: Scientificness, Credibility, and Belief in Pseudoscience," *Educational Psychology* 40, no. 7 (2020): 820–37.

26. YourDictionary Staff, "Examples of Pseudoscience in Different Fields," YourDictionary.com, September 13, 2022, https://examples.yourdictionary.com/examples-of-pseudoscience.html.

27. M. Morgan, W. B. Collins, G. G. Sparks, and J. R. Welch, "Identifying Relevant Anti-science Perceptions to Improve Science-Based Communication: The Negative Perceptions of Science Scale," *Social Sciences* 7, no. 64 (2018).

28. M. Carmen Erviti, M. Codina, and B. León, "Pro-science, Anti-science and Neutral Science in Online Videos on Climate Change, Vaccines and Nanotechnology," *Media and Communication* 8, no. 2 (2020): 329–38.

29. S. Blancke, M. Boudry, and J. Braeckman, "Reasonable Irrationality: The Role of Reasons in the Diffusion of Pseudoscience," *Journal of Cognition and Culture* 19, no. 5 (2019): 432–49.

30. A. Salonen, A. Hartikainen-Ahia, J. Hense, A. Scheersoi, and T. Keinonen, "Secondary School Students' Perceptions of Working Life Skills in Science-Related Careers," *International Journal of Science Education* 39, no. 10 (2017): 1339–52.

31. M. D. Gordin, "The Problem with Pseudoscience," *EMBO Reports* 18 (2017): 1482–85.

32. K. Tolley, "School Vaccination Wars: The Rise of Anti-science in the American Anti-vaccination Societies, 1879–1929," *History of Education Quarterly* 59, no. 2 (2019): 161–94, https://doi.org/10.1017/heq.2019.3.

33. S. O. Hansson, "Science Denial as a Form of Pseudoscience," *Studies in History and Philosophy of Science Part A* 63 (2017): 39–47; T. T. Desta and T. Mulugeta, "Living with COVID-19-Triggered Pseudoscience and Conspiracies," *International Journal of Public Health* 65 (2020): 713–14, https://doi.org/10.1007/s00038-020-01412-4.

34. D. Metin, J. Cakiroglu, and G. Leblebicioglu, "Perceptions of Eighth Graders Concerning the Aim, Effectiveness, and Scientific Basis of Pseudoscience: The Case of Crystal Healing," *Research in Science Education* 50 (2020): 175–202.

35. A. Nguyen and D. Catalan-Matamoros, "Digital Mis/disinformation and Public Engagement with Health and Science Controversies: Fresh Perspectives from Covid-19," *Media and Communication* 8, no. 2 (2020): 323–28 https://doi.org/10.17645/mac.v8i2.3352.

36. Monique Brouillette and Rebecca Renner, "Why Misinformation about COVID-19 Keeps Going Viral," *National Geographic*, September 22, 2020, https://www.nationalgeographic.co.uk/science-and-technology/2020/09/why-misinformation-about-covid-19s-origins-keeps-going-viral.

37. D. Dickson, "The Case for a 'Deficit Model' of Science Communication," Science and Development Network, June 24, 2005, https://www.scidev.net/global/editorials/the-case-for-a-deficit-model-of-science-communic/.

38. Tolley, "School Vaccination Wars."

39. I. Ruiz-Mallén, S. Gallois, and M. Heras, "From White Lab Coats and Crazy Hair to Actual Scientists: Exploring the Impact of Researcher Interaction and Performing Arts on Students' Perceptions and Motivation for Science," *Science Communication* 40, no. 6 (2018): 749–77.

Bibliography

Akos, P., and J. Kretchmar. "Gender and Ethnic Bias in Letters of Recommendation: Considerations for School Counselors." *Professional School Counseling* 20, no. 1 (2016): 102–13.

Arbesman, S. "Quantifying the Ease of Scientific Discovery." *Scientometrics* 86, no. 2 (2011): 245–50.

Astley, J., and L. J. Francis. "Promoting Positive Attitudes towards Science and Religion among Sixth-Form Pupils: Dealing with Scientism and Creationism." *British Journal of Religious Education* 32 (2010): 189–200.

Benderly, B. L. "A Trend toward Transparency for Ph.D. Career Outcomes?" *Science*, March 7, 2018. https://www.science.org/content/article/trend-toward-transparency-phd-career-outcomes.

Bettencourt, L. M. A., D. I. Kaiser, and J. Kaur. "Scientific Discovery and Topological Transitions in Collaboration Networks." *Journal of Informetrics* 3, no. 3 (2009): 210–21.

Blancke, S., M. Boudry, and J. Braeckman. "Reasonable Irrationality: The Role of Reasons in the Diffusion of Pseudoscience." *Journal of Cognition and Culture* 19, no. 5 (2019): 432–49.

Bloch, C., and M. P. Sørensen. "The Size of Research Funding: Trends and Implications." *Science and Public Policy* 42, no. 1 (2015): 30–43.

Bloodhart, B., M. M. Balgopal, A. M. A. Casper, L. B. Sample McMeeking, and E. V. Fischer. "Outperforming yet Undervalued: Undergraduate Women in STEM." *PLOS One* 15, no. 6 (2020): e0234685. https://doi.org/10.1371/journal.pone.0234685.

Bloom, N., C. I. Jones, J. Van Reenen, and M. Webb. "Are Ideas Getting Harder to Find?" *American Economic Review* 110, no. 4 (April 2020): 1104–44.

Blue, G. "Scientism: A Problem at the Heart of Formal Public Engagement with Climate Change." *ACME* 17, no. 2 (2018): 544–60.

Bornmann, L., and H. D. Daniel. "What Do We Know about the *h* Index?" *Journal of the American Society for Information Science and Technology* 58, no. 9 (2007): 1381–85.

Bruce, A. B. "Frankenfish or Fish to Feed the World? Scientism and Biotechnology Regulatory Policy." *Rural Sociology* 82, no. 4 (2017): 628–63.

Brue, K. "Work-Life Balance for Women in STEM Leadership." *Journal of Leadership Education* 18 (2019): 32–52.

Bullock, O. M., D. Colón Amill, H. C. Shulman, and G. N. Dixon. "Jargon as a Barrier to Effective Science Communication: Evidence from Metacognition." *Public Understanding of Science* 28, no. 7 (2019): 845–53.

Cañibano, C., P. D'Este, F. J. Otamendi, and R. Woolley. "Scientific Careers and the Mobility of European Researchers: An Analysis of International Mobility by Career Stage." *Higher Education* 80 (2020): 1175–93.

Canning, E. A., K. Muenks, D. J. Green, and M. C. Murphy. "STEM Faculty Who Believe Ability Is Fixed Have Larger Racial Achievement Gaps and Inspire Less Student Motivation in Their Classes." *Science Advances* 5, no. 2 (2019).

Carmen Erviti, M., M. Codina, and B. León. "Pro-science, Anti-science and Neutral Science in Online Videos on Climate Change, Vaccines and Nanotechnology." *Media and Communication* 8, no. 2 (2020): 329–38.

Chapman, C. A., et al. "Games Academics Play and Their Consequences: How Authorship, *h*-Index and Journal Impact Factors Are Shaping the Future of Academia." *Proceedings of the Royal Society B: Biological Sciences* 286 (2019): 1–9.

Copeland, S. "On Serendipity in Science: Discovery at the Intersection of Chance and Wisdom." *Synthese* 196, no. 6 (2019): 2385–2406.

Copur-Gencturk, Y., J. R. Cimpian, S. T. Lubienski, and I. Thacker. "Teachers' Bias against the Mathematical Ability of Female, Black, and Hispanic Students." *Educational Researcher* 49, no. 1 (2020): 30–43.

Cruz-Castro, L., and L. Sanz-Menéndez. "Mobility versus Job Stability: Assessing Tenure and Productivity Outcomes." *Research Policy* 39, no. 1 (2010): 27–38.

Danbold, F., and Y. J. Huo. "Men's Defense of Their Prototypicality Undermines the Success of Women in STEM Initiatives." *Journal of Experimental Social Psychology* 72 (2017): 57–66.

Desta, T. T., and T. Mulugeta. "Living with COVID-19-Triggered Pseudoscience and Conspiracies." *International Journal of Public Health* 65 (2020): 713–14. https://doi.org/10.1007/s00038-020-01412-4.

Dewsbury, B. M., C. Taylor, A. Reid, and C. Viamonte. "Career Choice among First-Generation, Minority STEM College Students." *Journal of Microbiology & Biology Education* 20, no. 3 (2019).

Dickson, D. "The Case for a 'Deficit Model' of Science Communication." Science and Development Network, June 24, 2005. https://www.scidev.net/global/editorials/the-case-for-a-deficit-model-of-science-communic/.

Dua, M. "Scientific Discovery and Its Rationality: Michael Polanyi's Epistemological Exposition." *Foundations of Science* 25, no. 3 (2020): 507–18.

Ecklund, E. H., S. A. James, and A. E. Lincoln. "How Academic Biologists and Physicists View Science Outreach." *PLOS One* 7, no. 5 (2012): e36240. https://doi.org/10.1371/journal.pone.0036240.

Ecklund, E. H., and J. Z. Park. "Conflict between Religion and Science among Academic Scientists?" *Journal for the Scientific Study of Religion* 48, no. 2 (2009): 276–92.

Estrada, M., et al. "Improving Underrepresented Minority Student Persistence in STEM." *CBE—Life Sciences Education* 15, no. 3 (2016).

Evans, J. H. "Epistemological and Moral Conflict between Religion and Science." *Journal for the Scientific Study of Religion* 50, no. 4 (2011): 707–27.

Falk, S. *The Light Ages: The Surprising Story of Medieval Science.* New York: W. W. Norton, 2020.

Fanelli, D. "How Many Scientists Fabricate and Falsify Research? A Systematic Review and Meta-analysis of Survey Data." *PLOS One* 4, no. 5 (2009): e5738. https://doi.org/10.1371/journal.pone.0005738.

Feibelman, Peter J. *A PhD Is Not Enough!* New York: Basic Books, 2011.

Ferngren, Gary B., ed. *Science and Religion: A Historical Introduction.* 2nd ed. Baltimore, MD: Johns Hopkins University Press, 2017.

Fisher, A. J., et al. "Structure and Belonging: Pathways to Success for Underrepresented Minority and Women PhD Students in STEM Fields." *PLOS One* 14, no. 1 (2019): e0209279. https://doi.org/10.1371/journal.pone.0209279.

Fournier, A. M. V., A. J. Holford, A. L. Bond, and M. A. Leighton. "Unpaid Work and Access to Science Professions." *PLOS One* 14 (2019). https://doi.org/10.1371/journal.pone.0217032.

Ganley, C. M., C. E. George, J. R. Cimpian, and M. B. Makowski. "Gender Equity in College Majors: Looking Beyond the STEM/Non-STEM Dichotomy for Answers Regarding Female Participation." *American Educational Research Journal* 55, no. 3 (2018): 453–87.

Gare, A. "Natural Philosophy and the Sciences: Challenging Science's Tunnel Vision." *Philosophies* 3, no. 4 (2018).

Garr-Schultz, A., and W. L. Gardner. "Strategic Self-Presentation of Women in STEM." *Social Sciences* 7, no. 2 (2018): article 20.

Gasparatou, R. "Scientism and Scientific Thinking." *Science & Education* 26 (2017): 799–812.

Gemme, B., and Y. Gingras. "Academic Careers for Graduate Students: A Strong Attractor in a Changed Environment." *Higher Education* 63 (2012): 667–83.

Goldacre, Ben. *Bad Science: Quacks, Hacks, and Big Pharma Flacks*. New York: Farrar, Straus & Giroux, 2011.

Gordin, M. D. "The Problem with Pseudoscience." *EMBO Reports* 18 (2017): 1482–85.

Gottlieb, J. J. "STEM Career Aspirations in Black, Hispanic, and White Ninth-Grade Students." *Journal of Research in Science Teaching* 55, no. 2 (2018): 1365–92.

Gould, J. "Planning a Postdoc before Moving to Industry? Think Again." *Nature*, December 3, 2020. https://www.nature.com/articles/d41586-020-03109-3.

Hansson, S. O. "Dealing with Climate Science Denialism: Experiences from Confrontations with Other Forms of Pseudoscience." *Climate Policy* 18, no. 9 (2018): 1094–1102.

———. "Science Denial as a Form of Pseudoscience." *Studies in History and Philosophy of Science Part A* 63 (2017): 39–47

Harris, C. M. "Quitting Science: Factors That Influence Exit from the STEM Workforce." *Journal of Women and Minorities in Science and Engineering* 25, no. 2 (2019): 93–118.

Hayter, C. S., and M. A. Parker. "Factors That Influence the Transition of University Postdocs to Non-academic Scientific Careers: An Exploratory Study." *Research Policy* 48, no. 3 (2019): 556–70.

Head, M. L., L. Holman, R. Lanfear, A. T. Kahn, and M. D. Jennions. "The Extent and Consequences of *P*-Hacking in Science." *PLOS Biology* 13, no. 3 (2015): e1002106. https://doi.org/10.1371/journal.pbio.1002106.

Hoenen, S., and C. Kolympiris. "The Value of Insiders as Mentors: Evidence from the Effects of NSF Rotators on Early-Career Scientists." *Review of Economic Statistics* 102, no. 5 (2020): 852–66.

Hoppe, T. A., et al. "Topic Choice Contributes to the Lower Rate of NIH Awards to African-American/Black Scientists." *Science Advances* 5, no. 10 (2019).

Jay, P. "The Humanities Crisis Then and Now." In *The Humanities "Crisis" and the Future of Literary Studies*, 7–31. New York: Palgrave Macmillan, 2014.

Kaufman, B. E. "The Real Problem: The Deadly Combination of Psychologisation, Scientism, and Normative Promotionalism Takes Strategic Human Resource Management down a 30-Year Dead End." *Human Resource Management Journal* 30, no. 1 (2020): 49–72.

Kell, D. B. "Scientific Discovery as a Combinatorial Optimisation Problem: How Best to Navigate the Landscape of Possible Experiments?" *BioEssays* 34, no. 3 (2012): 236–44.

Kotcher, J. E., T. A. Myers, E. K. Vraga, N. Stenhouse, and E. W. Maibach. "Does Engagement in Advocacy Hurt the Credibility of Scientists? Results from a Randomized National Survey Experiment." *Environmental Communication* 11, no. 3 (2017): 415–29.

Kuchynka, S. L., et al. "Hostile and Benevolent Sexism and College Women's STEM Outcomes." *Psychology of Women Quarterly* 42, no. 1 (2018): 72–87.

Laplane, L., et al. "Why Science Needs Philosophy." *PNAS* 116, no. 10 (2019): 3948–52. https://doi.org/10.1073/pnas.1900357116.

Larimore, R. A. "Preschool Science Education: A Vision for the Future." *Early Childhood Education Journal* 48 (2020): 703–14. https://doi.org/10.1007/s10643-020-01033-9.

Larkin, D. B. "Attending to the Public Understanding of Science Education: A Response to Furtak and Penuel." *Science Education* 103, no. 5 (2019): 1294–1300.

Larson, R. C., N. Ghaffarzadegan, and M. G. Diaz. "Magnified Effects of Changes in NIH Research Funding Levels." *Service Science* 4, no. 4 (2012): 382–95.

Larson, R. C., N. Ghaffarzadegan, and Y. Xue. "Too Many PhD Graduates or Too Few Academic Job Openings: The Basic Reproductive Number R0 in Academia." *Systematic Research and Behavioral Science* 31, no. 6 (2014): 745–50.

Laudel, G., and J. Gläser. "From Apprentice to Colleague: The Metamorphosis of Early Career Researchers." *Higher Education* 55 (2008): 387–406.

Laursen, S., C. Liston, H. Thiry, and J. Graf. "What Good Is a Scientist in the Classroom? Participant Outcomes and Program Design Features for a Short-Duration Science Outreach Intervention in K–12 Classrooms." *CBE Life Sciences Education* 6, no. 1 (2007): 49–64.

Leslie, S.-J., A. Cimpian, M. Meyer, and E. Freeland. "Expectations of Brilliance Underlie Gender Distributions across Academic Disciplines." *Science* 347 (2015): 262–65.

Leslie, T. "'I Decided': Listening to the Voices of Academically Resilient Black and Hispanic Students in College STEM Programs." *Dissertation Abstracts International Section A: Humanities and Social Sciences* 77 (2017).

Li, J., Y. Yin, S. Fortunato, and D. Wang. "Scientific Elite Revisited: Patterns of Productivity, Collaboration, Authorship and Impact." *Journal of the Royal Society Interface* 17, no. 165 (2020). http://doi.org/10.1098/rsif.2020.0135.

Lindholm, M. "Promoting Curiosity? Possibilities and Pitfalls in Science Education." *Science & Education* 27, nos. 9–10 (2018): 987–1002.

Lovitts, B. E., and C. Nelson, "The Hidden Crisis in Graduate Education: Attrition from Ph.D. Programs." *Academe* 86, no. 6 (2000): 44–50.

Markowitz, D. M., and J. T. Hancock. "Linguistic Obfuscation in Fraudulent Science." *Journal of Languages and Social Psychology* 35, no. 4 (2016): 435–45.

Martin, B. "Scientific Fraud and the Power Structure of Science." *Prometheus* 10, no. 1 (1992): 83–98.

Martin, D. C. "It's Not My Party: A Critical Analysis of Women and Minority Opposition towards STEM." *Critical Questions in Education* 7, no. 2 (2016): 96–115.

Martinez-Conde, S. "Has Contemporary Academia Outgrown the Carl Sagan Effect?" *Journal of Neuroscience* 36, no. 7 (2016): 2077–82. https://doi.org/10.1523/JNEUROSCI.0086-16.2016.

Matheson, A. "Five Steps for Structural Reform in Clinical Cancer Research." *American Journal of Public Health* 100, no. 4 (2010): 596–603.

Mayes, C. R., and D. B. Thompson. "What Should We Eat? Biopolitics, Ethics, and Nutritional Scientism." *Journal of Bioethical Inquiry* 12, no. 4 (2015): 587–99.

McClain, C. "Practices and Promises of Facebook for Science Outreach: Becoming a 'Nerd of Trust.'" *PLOS Biology* 15, no. 6 (2017): e2002020. https://doi.org/10.1371/journal.pbio.2002020.

McClain, C., and L. Neeley. "A Critical Evaluation of Science Outreach via Social Media: Its Role and Impact on Scientists." *F1000Research* 3 (2015). https://doi.org/10.12688/f1000research.5918.1.

McCullough, L. "Proportions of Women in STEM Leadership in the Academy in the USA." *Educational Sciences* 10 , no. 1 (2020): 1–13.

Meador, A. "Examining Recruitment and Retention Factors for Minority STEM Majors through a Stereotype Threat Lens." *School Science & Mathematics* 118, nos. 1–2 (2018): 61–69.

Mervis, J. "Data Check: U.S. Government Share of Basic Research Funding Falls Below 50%." *Science*, March 9, 2017. https://www.science.org/content/article/data-check-us-government-share-basic-research-funding-falls-below-50.

Metin, D., J. Cakiroglu, and G. Leblebicioglu. "Perceptions of Eighth Graders Concerning the Aim, Effectiveness, and Scientific Basis of Pseudoscience: The Case of Crystal Healing." *Research in Science Education* 50 (2020): 175–202.

Milojevic, S., F. Radicchi, and J. P. Walsh. "Changing Demographics of Scientific Careers: The Rise of the Temporary Workforce." *PNAS* 115, no. 50 (2018): 12616–23.

Moore, A. "Bad Science in the Headlines. Who Takes Responsibility When Science Is Distorted in the Mass Media?" *EMBO Reports* 7, no. 12 (2006): 1193–96.

Morgan, M., W. B. Collins, G. G. Sparks, and J. R. Welch. "Identifying Relevant Anti-science Perceptions to Improve Science-Based Communication: The Negative Perceptions of Science Scale." *Social Sciences* 7, no. 64 (2018).

Mullin, Rick. "Behind the Scenes at the STEM-Humanities Culture War." *C&EN* 97 (2019): 26–29.

Nadkarni, N. M., and A. E. Stasch. "How Broad Are Our Broader Impacts? An Analysis of the National Science Foundation's Ecosystem Studies Program and the Broader Impacts Requirement." *Frontiers in Ecology and the Environment* 11, no. 1 (2013): 13–19.

National Academies Press. *The Science of Effective Mentorship in STEMM.* Washington, DC: National Academies Press, 2019. doi:10.17226/25568.

Nealy, S., and M. K. Orgill. "Postsecondary Underrepresented Minority STEM Students' Perceptions of Their Science Identity." *Journal of Negro Education* 88, no. 3 (2019): 249–68.

Nguyen, A., and D. Catalan-Matamoros. "Digital Mis/disinformation and Public Engagement with Health and Science Controversies: Fresh Perspectives from Covid-19." *Media and Communication* 8, no. 2 (2020): 323–28. https://doi.org/10.17645/mac.v8i2.3352.

Nowogrodzki, A. "Most US Professors Are Trained at Same Few Elite Universities." *Nature*, September 21, 2022. https://www.nature.com/articles/d41586-022-02998-w.

Ohlers, C. "The 'Conflict Thesis' of Science and Religion: A Nexus of Philosophy of Science, Metaphysics, and Philosophy of Religion." *Journal of Biblical and Theological Studies* 2, no. 2 (2017): 208–33.

Parker, J. N., C. Lortie, and S. Allesina. "Characterizing a Scientific Elite: The Social Characteristics of the Most Highly Cited Scientists in Environmental Science and Ecology." *Scientometrics* 85 (2010): 129–43

Parsons, E. C. M. "'Advocacy' and 'Activism' Are Not Dirty Words—How Activists Can Better Help Conservation Scientists." *Frontiers in Marine Science* 3, no. 229 (2016).

Peels, R. "Ten Reasons to Embrace Scientism." *Studies in History and Philosophy of Science Part A* 63 (2017): 11–21.

Petersen, A. M. "Quantifying the Impact of Weak, Strong, and Super Ties in Scientific Careers." *PNAS* 112, no. 34 (2015): E4671–80.

Pielke, R. A. *The Honest Broker: Making Sense of Science in Policy and Politics.* Cambridge: Cambridge University Press, 2007.

———. "When Scientists Politicize Science: Making Sense of Controversy over *The Skeptical Environmentalist.*" *Environmental Science & Policy* 7 (2004): 405–17.

Ramsey, L. R., D. E. Betz, and D. Sekaquaptewa. "The Effects of an Academic Environment Intervention on Science Identification among Women in STEM." *Social Psychology of Education* 16, no. 3 (2013): 377–97.

Reid, L. "Scientism in Medical Education and the Improvement of Medical Care: Opioids, Competencies, and Social Accountability." *Health Care Analysis* 26, no. 2 (2018): 155–70.

Roberts, M. R. "Realizing Societal Benefit from Academic Research: Analysis of the National Science Foundation's Broader Impacts Criterion." *Social Epistemology* 23, nos. 3–4 (2009): 199–219.

Robnett, R. D., and S. E. Thoman. "STEM Success Expectancies and Achievement among Women in STEM Majors." *Journal of Applied Developmental Psychology* 52 (2017): 91–100.

Rose, N. A., and E. C. M. Parsons. "'Back Off, Man, I'm a Scientist!': When Marine Conservation Science Meets Policy." *Ocean & Coastal Management* 115 (2015): 71–76.

Ruben, A. "Scientists Should Defend, Not Defund, the Humanities." *Science*, September 23, 2015. https://www.science.org/content/article/scientists-should-defend-not-defund-humanities.

Ruiz-Mallén, I., S. Gallois, and M. Heras. "From White Lab Coats and Crazy Hair to Actual Scientists: Exploring the Impact of Researcher Interaction and Performing Arts on Students' Perceptions and Motivation for Science." *Science Communication* 40, no. 6 (2018): 749–77.

Rupert, D. D., A. C. Nowlan, O. H. Tam, and M. G. Hammell. "Ten Simple Rules for Running a Successful Women-in-STEM Organization on an Academic Campus." *PLOS Computational Biology* 16, no. 5 (2020): e1007754. https://doi.org/10.1371/journal.pcbi.1007754.

Rutjens, B. T., R. M. Sutton, and R. van der Lee. "Not All Skepticism Is Equal: Exploring the Ideological Antecedents of Science Acceptance and Rejection." *Personality & Social Psychology Bulletin* 44, no. 3 (2018): 384–405.

Ryan, A. M., D. D. King, F. Elizondo, and P. Wadlington. "Social Identity Management Strategies of Women in STEM Fields." *Journal of Occupational & Organizational Psychology* 93, no. 2 (2020): 245–72.

Rzhetsky, A., J. G. Foster, I. T. Foster, and J. A. Evans. "Choosing Experiments to Accelerate Collective Discovery." *PNAS* 112, no. 47 (2015): 14569–74.

Saffie-Robertson, M. C. "It's Not You, It's Me: An Exploration of Mentoring Experiences for Women in STEM." *Sex Roles* 83, no. 4 (2020): 533–79.

Salonen, A., A. Hartikainen-Ahia, J. Hense, A. Scheersoi, and T. Keinonen. "Secondary School Students' Perceptions of Working Life Skills in Science-Related Careers." *International Journal of Science Education* 39, no. 10 (2017): 1339–52.

Santos, L. F. "The Role of Critical Thinking in Science Education." *Journal of Education and Practice* 8, no. 20 (2017): 159–73

Sass, T. R. "Understanding the STEM Pipeline." National Center for Analysis of Longitudinal Data in Education Research, Calder Working Paper 125, January 2015. https://files.eric.ed.gov/fulltext/ED560681.pdf.

Saw, G., C. N. Chang, and H. Y. Chan. "Cross-Sectional and Longitudinal Disparities in STEM Career Aspirations at the Intersection of Gender, Race/Ethnicity, and Socioeconomic Status." *Educational Researcher* 47, no. 8 (2018): 525–31.

Scales, D., S. Gorman, and J. Gorman. "Resist Pseudoscience with Respect, Not Ridicule." *Nature*, June 2, 2020. https://www.nature.com/articles/d41586-020-01626-9.

Schmid-Petri, H. "Politicization of Science: How Climate Change Skeptics Use Experts and Scientific Evidence in Their Online Communication." *Climatic Change* 145, no. 3 (2017): 523–37.

Sedlak, D. "Crossing the Imaginary Line." *Environmental Science & Technology* 50, no. 18 (2016): 9803–4.

Shulman, H. C., G. N. Dixon, O. M. Bullock, and D. Colón Amill. "The Effects of Jargon on Processing Fluency, Self-Perceptions, and Scientific Engagement." *Journal of Language and Social Psychology* 39, nos. 5–6 (2020): 579–97.

Smith-Doerr, L. "Hidden Injustice and Anti-science." *Engaging Science, Technology, and Society* 6 (2020): 94–101.

Sosa, B., E. Fontans-Álvarez, D. Romero, A. da Fonseca, and M. Achkar. "Analysis of Scientific Production on Glyphosate: An Example of Politicization of Science." *Science of the Total Environment* 681 (2019): 541–50.

Stenmark, M. "What Is Scientism?" *Religious Studies* 33 (1997): 15–32.

Stephan, P. *How Economics Shapes Science.* Cambridge, MA: Harvard University Press, 2012. doi:10.4159/harvard.9780674062757.

Stroebe, W., T. Postmes, and R. Spears. "Scientific Misconduct and the Myth of Self-Correction in Science." *Perspectives on Psychological Science* 7, no. 6 (2012): 670–88.

Teicholz, N. "The Scientific Report Guiding the US Dietary Guidelines: Is It Scientific?" *BMJ* 351 (2015).

Tesler, M. "Elite Domination of Public Doubts about Climate Change (Not Evolution)." *Political Communication* 35, no. 2 (2018): 306–26.

Thomas, N., J. Bystydzienski, and A. Desai. "Changing Institutional Culture through Peer Mentoring of Women STEM Faculty." *Innovative Higher Education* 40, no. 2 (2015): 143–57.

Thornhill, T. "We Want Black Students, Just Not You: How White Admissions Counselors Screen Black Prospective Students." *Sociology of Race and Ethnicity* 5 (2019): 456–70.

Tolley, K. "School Vaccination Wars: The Rise of Anti-science in the American Anti-vaccination Societies, 1879–1929." *History of Education Quarterly* 59, no. 2 (2019): 161–94. https://doi.org/10.1017/heq.2019.3.

Ungureanu, J. C. "'Your God Is Too Small': Retracing the Origins of Conflict between Science and Religion." *Theology and Science* 20, no. 1 (2022): 23–45.

van der Linden, S., A. Leiserowitz, S. Rosenthal, and E. Maibach. "Inoculating the Public against Misinformation about Climate Change." *Global Challenges* 1, no. 2 (2017): 1600008.

van Veelen, R., B. Derks, and M. D. Endedijk. "Double Trouble: How Being Outnumbered and Negatively Stereotyped Threatens Career

Outcomes of Women in STEM." *Frontiers in Psychology* 10 (2019). https://doi.org/10.3389/fpsyg.2019.00150.

Varner, J. "Scientific Outreach: Toward Effective Public Engagement with Biological Science." *BioScience* 64, no. 4 (2014): 333–40. https://doi.org/10.1093/biosci/biu021.

von Hippel, T., and C. von Hippel. "To Apply or Not to Apply: A Survey Analysis of Grant Writing Costs and Benefits." *PLOS One* 10 (2015): e0118494. https://doi.org/10.1371/journal.pone.0118494.

Walker, R., and P. R. da Silva. "Emerging Trends in Peer Review—a Survey." *Frontiers in Neuroscieince* 9 (2015). https://doi.org/10.3389/fnins.2015.00169.

Walsh, J. P., and Y. N. Lee. "The Bureaucratization of Science." *Research Policy* 44, no. 8 (2015): 1584–1600.

Williams-Watson, S. "Persistence among Minority STEM Majors: A Phenomenological Study." *ProQuest Dissertations and Theses* (2017).

Woolston, C. "PhDs: The Tortuous Truth." *Nature* 575 (2019): 403–6.

Yadav, A., et al. "The Forgotten Scholar: Underrepresented Minority Postdoc Experiences in STEM Fields." *Educational Studies* 56, no. 2 (2020): 160–85.

Zaboski, B. A., and D. J. Therriault. "Faking Science: Scientificness, Credibility, and Belief in Pseudoscience." *Educational Psychology* 40, no. 7 (2020): 820–37.

Index

About the Author

Paul M. Sutter is a theoretical cosmologist at the Institute for Advanced Computational Science at Stony Brook University and an external advisor for NASA's Innovative Advanced Concepts program. He is an award-winning science communicator, author of *Your Place in the Universe* and *How to Die in Space*, and host of several TV and digital series. He also writes and hosts his own popular *Ask a Spaceman* podcast. Paul is a globally recognized leader on the intersection of art and science and has served as a United States cultural ambassador. He lives in Connecticut with his wife and two children.